Technology Is Not the Problem

Timandra Harkness

H Q

ONE PLACE. MANY STORIES

HQ
An imprint of HarperCollins*Publishers* Ltd
1 London Bridge Street
London SE1 9GF

www.harpercollins.co.uk

HarperCollins*Publishers*
Macken House, 39/40 Mayor Street Upper
Dublin 1, D01 C9W8, Ireland
This edition 2025

1
First published in Great Britain by HQ,
an imprint of HarperCollins*Publishers* Ltd 2024

ISBN: 978-0-00-849464-3

Printed and bound in the UK using 100% Renewable
Electricity at CPI Group (UK) Ltd

For more information visit: www.harpercollins.co.uk/green

For Dennis,
From the rock of whose love I am shaping myself.

Contents

Introduction

This is not the book I was expecting to write.

I've spent over a decade writing and talking about technology, and how central it is to the ways our lives are changing. I am excited about the positive potential of technology to make human lives better: medical advances for longer, healthier lives; more plentiful, clean energy; faster, cheaper travel; wider horizons through our ability to instantly communicate and share the best of our global culture. Inevitably, though, I also spend a lot of time warning about the potential downsides, especially of the digital, data-driven technologies through which we live more and more of our lives. It is, after all, an asymmetric world, in which a few companies or organisations collect data about us all, aggregate and analyse that data, and use it for their own purposes.

I have specific concerns about the effects of this shift into a data-driven society. It has implications for our privacy, when almost anything we do, from texting our friends to walking around our own home, may be feeding data into somebody's algorithm without our knowledge. The fastest growing area of the tech market, pre-pandemic, was smart speakers,[1] devices such as Amazon's Alexa that we install in our own homes to listen for our requests, play us music or answer our questions, like attentive servants.

The timesaving, effort-reducing qualities of this all-pervasive technology have led some people to talk about 'Digital Athens'. Like the citizens of that first democracy, freed to practise politics and philosophy by the labour of slaves, we will be freed for higher things by AI taking on the chores of everyday life. I think the ubiquitous technology is more like digital *Wolf Hall*. Your house is full of servants, but who are they really working for? They fetch what you command, but their true loyalty is to other masters. And, as your interface with the world, they can also filter what comes to you. They can even nudge you into doing things without your conscious decision, whether that's something you would have chosen – going for a run, perhaps – or something you definitely would not have.

Technology also has effects on where we focus our time and attention, when there's always a new funny video, or a new response to our latest witty post to check out. Most of all, the ease of following where the algorithms lead puts us all in the passenger seat of our own lives. Human judgement, responsibility and initiative are handed over more and more to automated systems. Even when the systems are supposedly designed for our own good, I am deeply concerned that what it means to be an adult, a citizen, a *person*, is being eroded by our willingness to hand over control to technology.

But, as the book's title suggests, I found myself questioning whether technology is the problem, or whether the problem might simply be *us*. I became interested in one particular aspect of how technology affects our lives. It's a change so all-pervading that writing about it can feel like talking to my fellow fish about what it's like to live in water. It's hard even to remember a time when things were different, so let me give you an example.

When I first moved to London, way back in the previous century, I never went anywhere without my *A–Z*. It's a book of maps of

London streets, with the road names in an index in the back (hence, A to Z). Anywhere in London, I could look up the street I was on, and the street where I was going, and work out how to get there, following the map from page to page as I walked. I also carried a Tube map, so I could plan my journeys on the Underground, and sometimes a train timetable if I wanted to use overground trains. Bus routes were an arcane mystery, and knowing which bus went where marked you out as a proper Londoner.

I can't remember when I stopped carrying an A–Z. I have now lived here long enough that I know where a lot of places are, but the significant change is that I can access all that information, and more, on my mobile phone. All I need to do is put in a postcode or address, and an app will offer me a menu of ways to get there – walking, cycling, driving or public transport – tailored to my starting point. It will tell me when the buses are due, and whether there are delays on the Tube. I can follow a blue dot showing my location or, if I don't want to read a map, some apps just put an arrow on the screen showing me which way to go.

Instead of using the same information as everyone else, I now get only information pre-tailored to my needs. I don't even have to think about my route, I can just pick from the options offered. This is, increasingly, how we expect the world to work. We don't get the same experience as anyone else, online or offline, because we're not the same as anyone else.

You may be one of the few people who was startled to hear a television advert address them by name. 'Welcome, Rebecca, to the Unlimited,' said the presenter of Skinny Mobile's ad on New Zealand's TVNZ, but only to viewers called Rebecca. Streaming customers with one of the 200 most popular names all heard the presenter say theirs. The UK's Channel 4 has also experimented with advertising so personalised it says your name. It didn't go down well. Hearing our own name does grab our attention, as intended, but there's a big difference between unexpectedly hearing a friend call you across the street, and finding a stranger in your living room being too familiar. Viewers complained, comparing the experience to the sci-fi film *Minority Report,* in which Tom Cruise's character is greeted personally by every billboard he passes.

We don't like it when the attentive algorithms get too personal. But we also don't like it when our digital world isn't personalised enough; when we get adverts for the thing we've already bought, or for the wrong sports team. That ambivalence runs through our relationship with the technology that profiles and targets us, both online and, increasingly, in the physical world.

If you have a Smart TV, or watch programmes via the internet, your viewing is a two-way process that feeds news and entertainment to you, and data back the other way, about your viewing habits and other things that happen on your internet connection. When I visited their smart home hub in Surrey, Samsung told me that their system knows when the kids are upstairs and the adults are watching alone. They *could* target TV ads to each individual, if they thought we'd accept it. Google make policy decisions about how much to target television ads, based on the profile they have built on each of us. Among the services which can be removed from the menu are divorce lawyers. If your TV doesn't show you adverts that call you by your first name, it's probably because you don't like it.

In the physical world, too, customised products and services are proliferating. You can buy children's books in which the hero has not only your kid's name, but their skin, hair and eye colour. You can get an Ordnance Survey map centred on your house, pyjamas printed with your pet's face, or a customisable refrigerator – also from Samsung, so perhaps it greets you by name when you open the door?

This personalised world would amaze our great-grandparents. In their day, only the rich got bespoke products and personal services. Everyone else got mass-produced goods and mass culture. In a sweeping generalisation that may upset some historians, I like to contrast the personalised century in which we're living with the mass century that came before. When I talk about centuries, I don't literally mean hundred-year periods that flip over like a station clock at midnight on 31 December. The mass century was well under way before 1900, and the personalised century was already kicking in its womb by the 1960s. Like all historical change, the transformation of a mass society to a personalised one happened unevenly to different people in different places and is still happening today.

Intrigued by this cultural, economic and political transformation, I started looking into how technology makes it possible by collecting data and building profiles of each of us. Today, as we conduct so much of our lives either through, or at least constantly accompanied by, digital devices, there is no shortage of raw material. How do algorithms turn all that information into useful insights about what makes each of us unique? Can Big Data *really* know me better than I know myself?

Curious about claims that algorithms can map our personalities and predict our behaviour, I did an online psychological test. The University of Cambridge Psychometrics Centre asked me to specify, on a five-point scale, ranging from 'Strongly Disagree' to 'Strongly

Agree', how well certain statements apply to me: 'I get excited by new ideas'; 'I hold a grudge'; 'I am filled with doubts about things'; 'I believe that I am better than others'; 'You are walking in the desert, and you find a tortoise . . .' OK, not that last one; that was the 'Are you human?' test in *Blade Runner*.

Trying to answer the questions honestly made me aware of what I was feeling about taking the Psychometrics Centre test. It wasn't just intellectual curiosity; it was a small thrill that I was going to learn more about myself. Two things struck me: first, that I am an insufferable narcissist who never tires of hearing more about how unique and fascinating I am. I couldn't wait to get the results. I got a near-erotic hit of pleasure from reading that I am 79 per cent open to new experiences, 34 per cent conscientious, 66 per cent extravert, 50 per cent agreeable (workmates may disagree) and only 42 per cent neurotic, though I'm now worrying about whether I should be more conscientious, so make that 43 per cent.

Second, I realised how absurd I am to revel in my own unique-ness by letting a computer assign me a numerical value along just five measures, known as the Big Five psychological traits. Giving exact percentages makes it sound very scientific. I don't just love new and interesting things – I am 79 per cent open to novelty. But nobody can be that definite about my character from a hundred questions. Being only 34 per cent conscientious, and hence 66 per cent impulsive and spontaneous, I bashed through the test, giving answers that were sometimes contradictory. I think I said that I both was and wasn't good at finishing tasks. A lot of the answers depended on my mood at that moment.

The University of Cambridge Psychometrics Centre hasn't told me anything about myself that any of my friends couldn't have told me already. Will I do that kind of online test again anyway? You bet I will. I will also do those quizzes that promise to reveal which zoo animal, eighteenth-century philosopher or Disney villain I am (Ursula from *The Little Mermaid*, thanks for asking). I want to be recognised, if only by a machine, for the special person I feel myself to be.

After I had done their psychological test, I gave the Cambridge Psychometrics Centre access to my Twitter account so they could profile me from my tweets, and got markedly different results. According to their analysis of my social media activity, I am less conscientious, less agreeable, and positively introverted. I am also, according to their analysis of my Twitter feed, male (I'm not) and aged around thirty (I'm not), which does help to explain the adverts I get for watches and beard care products.

Predicting the Big Five personality traits from social media activity is not as accurate as you may have heard. If you panicked about voters who were tricked by the evil genius of psychometric microtargeting, you can relax a little: those techniques are nowhere near as powerful as companies like Cambridge Analytica claimed. But don't relax completely. You should still be worried, just not for the reasons that you may think.

The Cambridge Psychometrics Centre was not involved in political campaigns that tried to use psychological techniques to influence voter behaviour. But the original research using Facebook data, on which that later commercial work was based, was done at the centre in 2012 by Michal Kosinski and colleagues. According to their paper, 'Private traits and attributes are predictable from digital records of human behaviour'[2] – not just age and gender, but sexual orientation, intelligence, and the Big Five personality traits, could all be predicted from what people liked on Facebook. 'People may choose not to reveal certain pieces of information about their lives, such as their sexual orientation or age,' wrote the researchers, 'and yet this information might be predicted in a statistical sense from other aspects of their lives that they do reveal.'

Some of the findings were surprising. Liking curly fries on Facebook was found to predict higher intelligence, for example. But it's worth unpicking what the word 'predict' means in this context. In everyday language, predicting something means you think it will happen. You see a person standing by a puddle, and you predict that, when the approaching bus drives through the puddle, they will be soaked. That prediction is based on your previous observations of water and buses. If you were using a dating app, you might predict that, when you meet in person, your date will probably be shorter, older, fatter and poorer than their profile claimed. You cynic. That prediction is based on your previous experience of online dating, or what you've read about the most common lies people tell about themselves. Let's hope that you're proved wrong, and that this person is the exception that makes dating apps worthwhile.

Michal Kosinski and his co-researchers use the statistical sense of 'predict', which is closer to anticipating how an internet date will turn out than the horrible inevitability of getting soaked by a passing bus. There's no guarantee that any one person they tag as intelligent has that quality in real life, but they think it is more likely, statistically, that they will be intelligent when compared to a random person who expressed no love for curly fries. Not *much* more likely, but better than pure chance.

When statisticians talk about one quality predicting another, this is what they mean: they are comparing groups and saying how much more likely one group is to include what they're looking for. Suppose you're looking for people with beards in order to advertise your beard care products. You can save yourself a lot of money and time by omitting women and children from your target audience, because hardly any of them have beards. Not all men have beards, but you have improved your odds by including only them in your target audience.

None of the predictive measures Kosinski and his colleagues found were anything like as strong as 'man' predicting 'beard', which is why the claims made by Cambridge Analytica were wildly exaggerated. Facebook have also tightened up access to their users' data since 2012, making it much harder for researchers to draw conclusions about us from our feelings about curly fries. I'd love to think this was out of respect for our privacy, but it's more likely to be about monopolising and monetising valuable information. Facebook may not know you better than your family and friends, but it does have more data on you.

I'm ashamed to say that I used that phrase 'Big Data knows you better than you know yourself' to sell my first book. Of course, it's not true. Neither data nor the computer programs that analyse it can know *anybody* in the way one person knows another person. Having no mind of its own, a computer program can never imagine itself inside somebody else's mind. It can't empathise, or intuit why somebody said or did something. All it can do is measure, record, compare with previous data and assign a probability that some other data will also be observed.

That is why the Cambridge Psychometric Centre's algorithm assigns me a position well into the masculine side of the gender axis, and around the age of thirty. It's because others who previously participated in their research, and whose tweets resemble mine in some way, were male and aged around thirty, or said they were. This character portrait is almost the opposite of being intimately known and understood by another person. Humans have theory of mind. We use our imaginations to understand what it might be like to be someone else, and what we would do if we were them. The Psychometric Centre, by contrast, is engaged in a mass sorting exercise, looking for statistical relationships in large populations and then applying the results to each of us. On average, I am a thirty-year-old man.

That mathematical model of a person, constructed by automated systems, is my avatar in this personalised world. I didn't make it myself, but technology steers my life along tracks chosen for *it*, not me. Most options are removed without my ever knowing they existed, whether that's the dating profile of the man who might otherwise have become the father of my children, or the job that might have been my first step towards leading a technology company.

This is how technology creates our personalised world. Instead of me choosing the direction of my life and seeking out the opportunities, people or ideas that I think will help me on my way, I get to choose from a limited menu designed for my digital double. Unlike me, that persona cannot think for itself or decide to do anything I haven't done before. There is no *who* at the centre of my personalised world, only a *what*. The first casualty of personalisation is the person.

But this proxy personalisation wouldn't work at all if I didn't play my part. The more I looked into *how* technology creates this world of digital profiles and algorithmic targeting, the more I asked myself a new question: *why* are we living in this personalised world? Yes, it makes money for companies who can successfully predict what we want and what we'll do, and nudge us towards buying what we're offered. Yes, it's more convenient to use a travel app than to carry an *A–Z*, a Tube map and a train timetable. Ten years ago, I would also have said that most people have no idea how intrusive the technology is, and that, if they knew, they would opt out. Today, though, most people have at least a general idea that their online experience is shaped by data gathered from their own activity. Nevertheless, however much we complain about the creepiness, and adjust our device settings to block ads and cookies, we keep going back for more.

I had the growing feeling that something else is going on. Something about *us* is shaping the technology that, in turn, shapes our lives. But *what*? As I was asking myself this question, an unlikely thing happened that crystallised the answer starting to form in my mind. I was invited to Berlin, to talk to the leaders of some fashion companies about what was changing in their industry.

I am the least fashion-conscious person you're ever likely to meet. At the time, I was making a radio programme about the future of fashion. In New York, I interviewed a designer who was vocally angry with the terrible boring people who just wore black all the time. There was an awkward pause when the designer suddenly noticed that *I* was wearing a black jumper, black jeans and black trainers, before they very graciously said they liked my boxing boots. So why ask *me*, of all people, to talk to leading fashion company bosses? Because the conference organiser recognised that digital technology was transforming their business. The company leaders hadn't yet grasped that this was not landscape gardening, but tectonic plates shifting beneath their feet. My job was to spell it out for them.

Sometimes, to understand something, you need to look at it from the other side. I was used to talking to people about data: what it can do; how it works; the social and ethical pitfalls it brings. But to get that across to people who lived and breathed fashion, I needed to start from where they were: an industry that anticipates what people will want to say about themselves, and produces a range of options from which customers can choose, to express who they are.

In one way, profiling and data were disrupting their businesses. Top-down advertising is becoming less important than peer-to-peer influence, and brands need to engage directly with their audience via social media. The boundaries between shaping a brand from the top down and consumers reshaping it within their own, ever-changing language of social meaning are eroded in the digital world. Companies can learn in real time how their customers are making fashion their own, and adapt, almost in real time, to the changing meanings of their brand.

I realised that this world of personalising technology was just a continuation of fashion by other means. The fashion industry has always been powered by the human urge to say something about ourselves by how we appear to others. How we look has always mattered. By choosing to wear a boring black jumper, black jeans and trainers, I was saying that I was more interested in what people say than in clothes, which, in hindsight, was a rather rude choice for interviewing fashion designers. For the talk in Berlin, I made an effort and wore a striking red dress and ostentatious boots.

Today, we have more choice than ever before, and more plat-forms on which to present ourselves. The business models are new, but they're built on deep and ancient foundations of the human urge for self-expression. I was able to give the fashion industry a completely new perspective on their own world, and I left Berlin with a new perspective on mine. I had the answer to my question.

Why do we embrace intrusive data-collecting technology? Because we need to be recognised. We need to feel that we are seen by others as our authentic selves. We embrace this personalised world because it meets human needs that are emotional, psychological, even existential. I spend my professional life studying the trends that will shape our future, especially around technology. But when people ask me to name the big idea that will most influence Western societies in the next thirty years or so, it's not colonising Mars, extreme longevity or brain–computer interfaces. It's not even digital data or AI, though they will be everywhere, and we'll conduct most of our everyday lives through them. The idea that will do most to shape our immediate future is identity.

Identity is the dominant way we understand ourselves in the world today. It seems to meet our needs, both to be recognised for the unique person we are, and to be recognised by others as belonging to them, and with them. For most of history, rigid social expectations pressured people to play narrowly defined roles in public, to hide sides of themselves that were important to them but socially taboo. That's why the Pride slogan, 'Bring your whole self to work', had such resonance. Today, we are encouraged to play roles that feel authentic to us, bringing to a public stage elements of our personality that would once have been private, if not secret.

Singer Sam Smith expressed this very eloquently on Instagram in 2019: 'Today is a good day so here goes. I've decided I am changing my pronouns to THEY/THEM ♥ after a lifetime of being at war with my gender I've decided to embrace myself for who I am, inside and out . . . I've been very nervous about announcing this because I care too much about what people think but fuck it! I understand there will be many mistakes and mis gendering but all I ask is you please please try. I hope you can see me like I see myself now. Thank you.'

You might be asking yourself why a successful, wealthy singer with millions of fans cares so much what pronouns other people use to talk about them. The *Sun* newspaper covered the story under the headline, 'Sam Smith splashes out on £12million London mansion after asking to be referred to as "they" rather than "he".'[3] But this would be missing the point. Sam Smith wants others to 'see me like I see myself now'.

As our freedom to choose how to live increases, our freedom to explore and express *who we are* expands with it. *What* we are, which social categories we fit into, is less of a barrier than ever before, so we might expect that question to matter less and less. Instead, identity has grown into a powerful but contradictory idea: it is both a unique, inner kernel, and an outward projection; both an essential, defining intersection of characteristics, and a performed role for which we write our own script. The word 'persona' comes from the mask used in Ancient Greek theatre, which projected the actor's voice and gave emphasis to his physical expressions. Now we each have an online persona that projects our identity into the world.

We can be whoever we want, but *who we are* is now the most important thing, from which everything else must follow. The problem is that our sense of self is fragile, driving us to want affirmation and reassurance from others. Why else would people – including me, obviously – care so much how strangers respond to them on social media?

It is wonderful that we have so much more choice than our ancestors about how to live our lives. That freedom, if we address it honestly and with courage, could give us a far stronger sense of who we are than any number of social media likes or followers. Ask yourself: on your deathbed, how will you want to look back at your own life? Won't you want your obituary to consist of more than just being seen the way you see yourself, online or offline? To say that you raised kind children perhaps, built houses that people lived in, got a PhD after years of work, or made an extraordinary garden out of an overgrown yard. That you founded a movement for political change, invented a new form of transport or composed a song that will be sung long after you are gone.

None of these things are yours for the asking, or at the twitch of a thumb on a smartphone screen. They demand effort, imagination and risk-taking. Unfortunately, although technology is not the root of the problem, it does make it ever easier to choose from a menu of pre-selected options, and harder and harder to find our own directions and build our own lives. Technology offers a constant menu of small choices and small social rewards, tiny affirmations that we are recognised, designed to keep us coming back for more. Our screens are more seductive than the pool in which Narcissus gazed at his own reflection in the Greek myth. We don't just get our online persona reflected back to us, in all its edited perfection. We get instant assurance in the form of likes, replies and personalised feeds that others also find our reflection irresistible.

This book sets out to answer four questions: where are we? How did we get here? Why does it matter? And where do we go from here?

I don't expect you to agree with everything I say in this book. I expect – in fact, I hope – that you will find it challenging. I look forward to many real-life arguments about the ideas you're about to read, and possibly to changing my mind about some of them.

Instead of pointing the finger at technology, I argue that the problem is you, and me, and all of us living in this digital hall of mirrors. We don't have a shrinking sense of who we are because of personalising technology. We have personalising technology because of our shrinking sense of who we are. The problem is not technology, but our own obsession with how others see us, and our insatiable need to be reassured that we are the person we want others to see. How to turn our energy away from the digital mirror of Narcissus, and out to the wider world, is the challenge this book tries to meet.

Although it doesn't address you by name, it is addressed to you as a person. Not just the person you are now, but the person you don't yet imagine you can become.

Chapter 1: From Mass Production to Customised Consumption

'Imagine a shop that only sells clothes suited to you.
Well, now you don't have to.'

The advert in my Twitter feed showed a man carrying his shopping in bags printed with his own face, along a street where every shop also had posters of his face. All the shops, and all the bags, were called Stephen, which presumably was his name.

Twitter, having inferred that I was a man, had targeted me accordingly with this advert for Thread, a shopping service. Curious, I pretended to be the man that Twitter assumed I was, gave Thread some data on my tastes and supposed physical shape, and then got regular emails from my personal stylist (assisted by AI, they told me) with a personalised menu of men's clothes to buy and wear. It lasted a couple of years, and I grew rather sorry for men, who seem to wear mainly blue jumpers that I couldn't tell apart. I did feel a tiny tingle of satisfaction when the email referred to my 'sophisticated, preppy style' even though it was my fictitious, male alter ego who had the penchant for chinos and deck shoes.

I feel ambivalent about targeted adverts. When they're too accurate, I feel I'm being stalked, even though I know the entire process is as automatic as a vending machine. When they're not accurate enough, I feel irrationally insulted that they've got me all wrong. Weirdly, I didn't mind that Twitter thought I was a man, but I was furious when they offered me products from the wrong football team.

I'm not alone in this ambivalence. Ten years ago, researchers found that four out of five consumers understood how their past online behaviour determines what adverts they see in future, were concerned about creepiness and erosion of privacy, but also thought they got benefits from seeing more relevant advertising.[4] In June 2023, UK consumers were equally likely to say they were annoyed by online ads based on their search histories and that they didn't mind seeing advertising in exchange for free content.[5] This book will dig into that ambivalence.

The way we buy things today is very different from a century ago. We have moved from a world of mass production and mass markets to a world of personalised consumption, and this chapter tells that story. In 1909, Henry Ford wrote, 'I will build a motor car for the great multitude. It will be so low in price that no man making a good salary will be unable to own one.'[6]

Ford's name became synonymous with the mass production of identical products, though in fact he was not the first to use the assembly line method. Two engineers in the UK built a steam-powered production line in Portsmouth to make vital components for Nelson's entirely wind-powered Navy, a century before Henry Ford adopted the assembly line system to manufacture the Model T.

In Ford's Detroit factory, the standard chassis moved along the line, teams of workers fitted interchangeable parts, and soon each car was built in an hour and a half, instead of the twelve hours it used to take.[7] The price of a Ford Model T fell from nearly $1,000 to under $300 by 1929, and the workers producing it earned well above average wages. This massive increase in efficiency and productivity put Ford's products in the reach of ordinary people. By 1930, there was one car on America's roads for every five Americans.[8]

Mass production continues today, but its high-tech capabilities can now combine interchangeable components in thousands or millions of permutations. You can order a new Ford car, not only in any colour you like (from the selection on the menu) but also with your choice of interior design and accessories, engine size, driver assist technology, and whether or not your steering wheel is heated;[9] 'Your Ford, Your Way', as the advert says.

Ford still use assembly lines, but they have reorganised their manufacturing process. Instead of producing large numbers of identical vehicles that sit around waiting to be sold, they produce cars to meet customer demand. Inviting customers to specify exactly what they want benefits Ford, because they have a guaranteed buyer who is willing to wait months for delivery, before they even begin to make the car.[10] Who wants to be part of 'a great multitude' when you can be a unique individual?

Personalised products are proliferating. My stepmother's dogs get bags of food with their names on, formulated for each of them according to their age, breed and lifestyle. After a telephone conversation between my stepmother and the company about Tipple the terrier getting a bit tubby, they sent a new formulation. It's a big change from the tubes of unidentifiable offal, bought from a market stall, that sustained previous dogs when I lived at home.

You can order personalised vitamins, formulated for your lifestyle and body type. You can send off a swab to get beauty products matched to your genes. I am sceptical that any of these are any better than a far cheaper, off-the-shelf version, but that's not really the point. We want to feel special and unique, and this bespoke approach to products adds value.

Our desire to stand out from the crowd may be a reaction against the mass-produced world of identical Ford cars, but the ease with which we can fulfil it is the fruit of that world. It's mass production, and the mass market, that gave us this almost unmanageable plethora of choice.

It was thinking about the ways in which the fashion industry is changing that brought home to me the transformation from a mass-produced world to one of personalised consumption.

What's your favourite item that you're wearing right now – and why? Maybe it has sentimental attachments – a present, an heirloom, or it reminds you of the holiday when you bought it. Perhaps it's just very comfortable – soft, or warm, or perfectly fitting. Maybe it makes you feel good in another way, because it's an outward expression of who you are. When you wear it, other people respond to you in the way you want. They see you as glamorous; capable; creative; as standing out from your surroundings or fitting perfectly into your social group.

We value the items we buy for reasons beyond how useful they are, or how much they cost to make, because we buy things for social, as well as practical reasons. We value what they signal to other people about us. That funny little ornament? Oh, just a fun thing you brought back when you trekked to Machu Picchu (because you're so well travelled and adventurous). That hand-woven hessian bag? You bought it in the farmers' market (because you're so environmentally responsible and you want to support local businesses). That band T-shirt? Oh, that was their first gig, before they were famous (because you're so ahead of the curve in music. Or you were, twenty years ago).

The more people can afford mass-produced goods, the less social value they have. Locally produced and artisanal products are desirable precisely because they can't be produced on a scale that would let everyone buy them. Centuries ago, when everything was handmade, everything was expensive, and most things were too expensive for most people. Today, though you may not be able to afford bespoke tailoring, you can find mass-produced things that say something special about you, thanks to the internet: a T-shirt with your favourite obscure philosopher's face, perhaps, or a poster of a cult film. If it doesn't already exist, technology can make it specially for you, at a price you can afford. Printing T-shirts to order is only the beginning of the technological change that is transforming the world of fashion.

A woman is dancing, in front of a packed stadium. Her sense of freedom, her confidence in her own body and the pleasure she's taking in the music shine in the looks she exchanges with the audience, and with the camera.[11] Her blonde hair swings across her tanned and toned shoulders. Her dance partner is about twice her height, and made of orange metal.

The dancer is Amy Purdy, Paralympic snowboarder, and she's dancing a duet with an industrial robot at the opening ceremony of the 2016 Paralympic Games in Rio. At one point the robot, Kuka, lifts her gently off the ground. Amy's curved, bladelike, carbon fibre feet tread the air gracefully. It's a poetic union of flesh and machine, human and technology, in joyous, life-affirming harmony.

But I am looking at her dress. A heavy, open material, a few shades lighter than Amy's own tanned skin, it softly hugs her figure, the wide skirt swinging with her hips to the rhythm of the music. It was created by Israeli designer Danit Peleg, whose work could revolutionise the world of fashion. 'It was super challenging,' Danit told me, 'because the event was in Brazil, Amy's in the US and I'm in Tel Aviv. And we all need to communicate on how this dress is going to look, and create something that fits her.'

To get all Amy's measurements, instead of using a tape measure, Danit asked Amy to take two photographs of herself. The Nettelo app turns data from a couple of selfies into a 3D body model, and over a hundred specific measurements. 'Just by clicking her shoulders,' said Danit, 'I was able to see how many centimetres we have between these two dots.' Adding Amy's data into the design software let Danit tailor the design to her measurements. 'And I created a dress that fits her like a glove without ever meeting in person.'

Anyone can download sewing patterns from the internet and print them, but when Danit talks about sending one of her designs to print, she means something much more radical. Amy's dress – not the pattern, but the actual garment she wore – was printed from a digital file sent across thousands of miles. Danit did go to Rio for the final fitting, but no changes were needed. 'We just tried it on her, it fits her perfectly, and I was able to go and see Rio for the rest of the few days.'

I first met Danit Peleg in 2018, while recording that programme about the future of fashion. Her small flat was in an old, thick-walled block, whitewashed walls reflecting the Tel Aviv sunshine, and unkempt houseplants filling every spare corner. In her studio, Danit showed me the entire manufacturing process, from first sketches to a finished garment that I could try on – a bomber jacket made in a desktop 3D printer.

I watched another jacket section emerge slowly from the 3D printer's nozzle. Just like an inkjet printer turning a digital file into a paper document, the print head travelled to and fro, depositing the melted, plastic-like material in exactly the right spots. Then it built up the layers until the frail lace of Danit's design was a robust lattice, flexible like knitted fabric, but holding its precise geometry as it stretched and bounced back. The effect is something like crochet, but springier; like lace, but heavier; like pierced leather, but lighter and more flexible.

If you like the sound of the bomber jacket, you can go to her website and order one. When I spoke with Danit on Zoom in 2021, she had one parcelled up and awaiting the courier who would take it to a customer she has never met. 'It's the first 3D printed garment you can buy on the internet,' Danit told me, 'or the only one, I'm not sure.' After a virtual fitting session with the same Nettelo app that Amy Purdy used, you can choose your colours and even have your name printed into the back – not embroidered afterwards, but as an integral part of the fabric.

'The way we consume our fashion will change quite a bit,' Danit told me. Her vision is that designs can be sold globally, as digital products, and the physical garments made locally. 'You can imagine small labs like that in every mall or every street, or in houses, eventually.' The result would be no more fabric offcuts and waste, because each section is printed to the exact shape and size required; no more shipping costs for containers full of new clothes; no leftover stock being sent to landfill.

In future, 'You will be able to choose the materials that you like,' said Danit, 'so if it's a summer day, you'll print the same dress from cotton, and if it's winter you'll print it with wool, and you'll have it ready within a few hours, or hopefully a few minutes, nearby. You buy your fibre spool from your favourite place, but you can buy the garment from someone else. You have a say in the production process.'

This all sounds surprisingly, nostalgically, familiar to me. I remember my mother buying patterns, fabric and buttons, and making clothes. She had skills I certainly didn't inherit with her sewing machine, which languishes at the bottom of my cupboard. Danit sees the digital world expanding the scope of fashion for self-expression, uniting the virtual world of social media and the physical world of wearing something that expresses who you are – or who you want to be in everyone else's eyes.

More than that, though, it changes the relationship between designer, manufacturer and buyer. 'What happens now,' as Danit told me, is that 'fashion designers design something, they send it to production, they put it in physical stores around the world and wait to see what happens with this garment, instead of asking the clients, "Do you like it?" before you produce it.'

Upmarket fashion designers have always made clothes to measure, at price tags most people could never afford. Now, mass-market fashion brands from Adidas to Zara are rethinking this process, and looking at the 'pull' dynamic of supplying what customers want. Instead of planning all their products many months ahead, fashion companies now monitor the trends emerging online and rapidly respond with new lines made as locally as possible.[12]

Soon, instead of waiting for the designer's new collection to hit the shops, you will try it on your digital body double in the virtual changing room, and order a new outfit to suit you, to fit you, and to arrive within a few minutes of where you are, before getting out of bed. 'No shipping or logistics. It's completely on demand,' said Danit. 'When I produce a garment, it's meant to be for someone. Every print is different. So it's really not a problem to create something that is custom made or personalised, made exactly for the customer.'

It is, as Nettelo put it, mass customisation.

*

We have always used our appearance as a way of communicating how we want others to see us. Archaeologists excavating a cave in Morocco found shell beads dating back to the middle Stone Age – between 140,000 and 150,000 years ago.[13] They showed signs of being coloured with ochre and threaded on strings, suggesting that they were worn by the Aterian people living in the area at the time. Anthropologist Steven L. Kuhn, part of the team who discovered the beads, thinks they played a symbolic as well as a decorative role for a growing population living in tribes or clans: 'They were probably part of the way people expressed their identity with their clothing,'[14] he said.

The clothes we wear are part of our social being. Uniforms exist to define a group. Whether they're official uniforms or the informal dress codes of friendship groups, workplaces and special occasions, they identify both those who belong, by wearing the uniform, and those who don't. The same leathers and boots in which I blend right in with other motorcyclists would make me stand out a mile at an expensive opera house.

There was a time, not so long ago, when I would have been ostracised for turning up in trousers, never mind leathers. The philosopher Elizabeth Anscombe is said to have been denied entry to a restaurant because they did not admit ladies wearing trousers. She removed the trousers.[15] Never pick an argument with a philosopher. This was in mid-twentieth-century America. In spite of a ruling by the US Attorney General in 1923 that women could wear trousers anywhere, women were not in practice allowed to wear trousers everywhere. Female senators could not wear trousers on the Senate floor until 1993.[16]

As late as 1978, two people in Chicago were convicted of violating a city law against wearing the clothing 'not belonging to his or her sex, with intent to conceal his or her sex', after being arrested leaving a diner. One was wearing a dress, nylon stockings, a wig, and a fur coat, the other a pants suit, high-heeled shoes and 'a bouffant hair style'. At the police station they were also discovered to be wearing suspender belts and bras.

They appealed to the Supreme Court, which reversed the conviction, not because choice of appearance was a fundamental right, but on the grounds that the defendants identified as women, and were preparing for gender reassignment surgery. The Court also noted that, 'The notion that the State can regulate one's personal appearance, unconfined by any constitutional strictures whatsoever, is fundamentally inconsistent with values of privacy, self-identity, autonomy, and personal integrity that Constitution was designed to protect.'[17]

This is a relatively new way of thinking about personal appearance. In most earlier societies (and still in many societies around the world) the idea that self-identity and autonomy mean the right to dress as you wish would be baffling. Clothing served practical purposes – to protect you from the weather, or from the physical challenges of your working life – and equally important social purposes, to signal who you were in a complex web of relationships.

The earliest sumptuary laws in England, passed in the fourteenth century, specified in great detail what the different social classes were allowed to wear. Only Lords could wear ermine (still the ceremonial wear of members of the British House of Lords). Nobles and richer merchants could wear precious metals, expensive cloth and other types of fur, according to their rank and wealth. Agricultural workers without property were only permitted to wear cheap cloth with belts made of rope.

It's telling, though, that these laws came in after the Black Death had cut a terrible swathe through the European population. Inherited rank in the feudal system was shaken up by merchants making money from trade, and farm workers demanding more pay and respect. The laws were made, in part, to try to keep in place a rigid hierarchy signified in outward appearance, a social structure that could no longer be taken for granted.

Similar laws were used in sixteenth-century England, and then in the new colonies in America, in another era of social structures in flux. A Massachusetts colonial law of 1651 forbids the wearing of gold or silver, lace and buttons, or silk hoods and scarves, to anyone with property worth less than £200 (except public servants or military officers).[18] The following year, Jonas Fairbanks was one of the people taken to court for wearing 'great boots' that came up over the knee, but escaped the ten shilling fine because he got the boots before the new law came in.[19]

Just as the booming wool trade and the sudden rise in agricultural wages allowed fourteenth century peasants and merchants to dress above their social rank, mass production has enabled even workers in routine jobs to buy many more clothes than one person can wear in a week. The internet gives shoppers in developed countries access to a far wider range than even the longest high street could offer, and Western democracies, at least, are much more tolerant of fashion choices that would have been outrageous even fifty years ago.

The importance of presenting to the world exactly the person you want others to see has also grown, and that isn't just about clothes. In the era of sharing everything from your home bookshelves to your breakfast through social media, almost everything you buy says something about who you are. Fifty years ago, the only people who would see your pans and kitchen chairs were your family and neighbours. Now every teaspoon is a potential extra in your Facebook or TikTok post: it's saying something about you. You could argue that this is because technology has made the division between public and private life more permeable, but the social media platforms don't force us to share every aspect of our home lives, our holidays, our baking and our babies. We *want* to put our lives on public display, through carefully composed shots and selected words, to build the self that we want to be in the minds of our online communities.

Today, we do far more of our shopping on the web than ever before. This was already the trend before Covid, and the combined effects of shops being closed and shoppers changing their routines accelerated it. But that's not the only aspect of fashion that's moving online.

The more our social lives happen through screens, the more important they become as the medium for us to show who we are. If all your friends can see you looking chic on Instagram or

TikTok, what does it matter if people on your street never see it? You probably don't even know them. Teenage girls may still pile into high-street shops with their friends, but they're also selecting outfits online, as recommended by algorithms, and art-directing the images they share on social media. They may not even buy the clothes to wear in real life; they can just take the selfie in the outfit and send it back.

In fact, why even put on a physical garment? You can now buy virtual fashion that only your digital self can wear. 'I really love this idea because firstly, it's environmentally friendly and secondly, clothing nowadays is more like an art form for social media,' influencer Daria Simonova told *Elle* magazine. 'Digital clothing is super convenient, and the design potential is huge because it's way cheaper.'[20]

All this digital activity generates data. From liking images of influencers wearing new styles, to searching and browsing online stores, to posting your own images, your online life is a stream of information for those with the tools to collect it. Your audience now includes companies who want to know what you do, and like, and buy, or might be persuaded to buy. Sitting in on every conversation, like the quiet friend-of-a-friend at your table who listens and never speaks but remembers every detail of who said what, is an algorithm.

The technology is new, but the desire to know more about us is at least a century old. Since the masses first had disposable incomes, companies have tried to learn more about their customers, partly to give them what they want, and partly to work out better ways of making them want something different.

George Gallup, founder of Gallup polling, had a PhD in

psychology. His dissertation, 'An Objective Method for Determining Reader Interest in the Content of a Newspaper',[21] took the sampling methods used by inspectors of wheat and water, and applied them to what people thought and felt about newspapers. Gallup headed the research department of advertising agency Young & Rubicam from 1932. This application of statistical methods to human minds proved very successful, on a population scale. Both Gallup and Henry Durant, founder of the British branch of Gallup, BIPO (British Institute of Public Opinion), made their reputations by correctly predicting election results.

Gallup predicted the victory of President Franklin D. Roosevelt in 1936, unlike his main competitor, the *Literary Digest* newspaper. Gallup foresaw that the newspaper's reader poll, though a much larger sample of voters, would over-represent people with cars and telephones, who were less likely to vote for Roosevelt. He weighted his much smaller sample to better represent American voters. Soon after their embarrassing defeat, the *Literary Digest* closed down.

Henry Durant, a researcher at the London School of Economics (LSE), predicted the 1945 election victory of Clement Attlee's Labour government. Durant, the son of a warehouseman, had got a scholarship to a good school and then worked as a clerk before becoming a sociologist. He described public opinion as 'a conventional yardstick which imparts to one person an opinion more or less equal in weight to the opinions of other individual persons'.

Durant saw opinion polling as a democratic project, a chance for ordinary people to get their voices heard between elections. 'Implicit in this arithmetical equalitarianism is the democratic premise that government is by, with and for the people,' he wrote in 1955. 'If you agree with the democratic premise of one man one vote, it logically follows that it is possible and helpful to collate and at the same time analyse a whole gamut of individual opinions.'[22] A few years later, he went still further, expressing the hope that combining sample surveys with computers would give a psychological X-ray of human minds.[23]

From the start, there has been enormous cross-fertilisation between market research for commercial purposes, social science research, and opinion polling by political parties, campaigns and governments. The rise of public opinion research answered a new need in the late nineteenth and early twentieth centuries. The new expansion of political power to the majority of people gaining the right to vote, and the expansion of their economic importance as people with money to spend, meant their motivations mattered like never before. Both politicians and advertisers wanted to know how people would behave, on a population scale, and were keen to use science and mathematics to find out.

Nielsen, the company that still measures and analyses media consumption and advertising impact, was founded by an engineer, Arthur C. Nielsen, in 1923. He started out testing conveyor belts and turbines, and then applied the same methods to consumer behaviour. Digital media still use the same statistical methods today, though much faster and at a greater scale.

The first advertising agencies, too, wanted to use the latest scientific methods on our minds and our money. They hoped to make adverts more effective and, just as important, to impress their own customers, the companies with products to sell.

Psychologist John B. Watson was on the payroll of American advertising firm J. Walter Thompson from 1920. Watson is credited as the founder of behaviourism, and his article, 'Psychology as the

Behaviorist Views It', published in the 1913 *Psychological Review*, is often called the behaviourist manifesto. He begins with a bold assertion: 'Psychology as the behaviorist views it is a purely objective experimental branch of natural science. Its theoretical goal is the prediction and control of behavior. Introspection forms no essential part of his methods, nor is the scientific value of its data dependent on the readiness with which they lend themselves to interpretation in terms of consciousness.'[24]

The behaviourist school saw humans, like other animals, as machines that would respond to the correct inputs with predictable outputs. It's easy to see how advertising, a field entirely concerned with understanding, predicting and controlling human behaviour, would welcome this approach. But behaviourism wasn't the only psychological theory applied to marketing.

In the 1920s, most men smoked cigarettes. Packs of cigarettes were sent to soldiers fighting in the trenches of the First World War. But women, certainly respectable women, did not. A woman was arrested in 1922 for smoking a cigarette on a New York City street.[25] This was an inconvenience for tobacco companies. If women took up smoking, they could double their market. So American Tobacco consulted Edward Bernays, widely called the founder of modern public relations, on how to get women smoking their Lucky Strikes.

Bernays was the nephew of Sigmund Freud, father of psychoanalysis. Like Freud, Bernays believed that emotions and the subconscious were powerful drivers of human behaviour, and he consulted psychoanalyst A. A. Brill about the deep drives that might make women want to smoke. Brill, who had studied with Freud, connected the 'phallic' cigarette with men's social potency, power and freedom.

On 31 March 1929, Easter Sunday, a group of fashionable young women walking on Fifth Avenue all lit up cigarettes or, as they called them, 'torches of freedom'. Newspapers covering the Easter Parade got a bonus story, with a whiff of scandal and an excuse to run photographs of attractive young women. 'Group of Girls Puff at Cigarettes as a Gesture of "Freedom"' said the front page of the *New York Times* the following day.[26]

The fact that the organiser also happened to be the secretary of one Edward Bernays was not mentioned. But smoking was now linked in the public mind with other contemporary moves towards women's emancipation – access to the vote, education and employment. There's no evidence that it drove a massive increase in women buying Lucky Strikes, but it's often cited as an example of Bernays' ambitious approach to public relations. Why just publish an advert when you can attempt to change the way everyone thinks about a product, or a cause, by making them feel differently about themselves?

Thirty years later, a car advert broke up glossy, colourful *Life* magazine with a mostly white page. 'Think Small' said the headline, under a tiny VW Beetle, printed in black on a blank background, a deliberate contrast with all the colourful, boastful, American car adverts at the time.

It was the start of an advertising campaign designed to appeal to people who felt they were different from the mass of customers buying large, luxurious cars. The adverts emphasised qualities the car had – reliability, simplicity – but just as importantly, they implied qualities that the *customer* had: Beetle owners rebel against the consumerist values of post-war America, say the ads. They choose simplicity, quirkiness and standing out from the crowd.

Perhaps ironically, this approach drove the VW Beetle to become the bestselling car of all time in 1972, overtaking the Ford Model T. Standing out from the crowd clearly had mass appeal in the 1960s. Though it ceased production in 2003, the classic Beetle remains one of the bestselling cars in history.

The study of how people behave, and what we want, has long been both a science and an art, and its changing methods reveal a lot about how human beings are regarded at any one time. Market researchers have used different techniques both to get acquainted with individuals, and to study overall trends. Today's data collection and analysis methods mean it's easier than ever before to combine the two, allowing the algorithms to build a model of you, personally, and the way you behave. This model is far from perfect, and far from personal, but on a population scale it has transformed marketing, and hence the market.

We are mostly unaware of the methods by which we are profiled and targeted, though we tend to have a general idea that our devices are observing us and snitching to advertisers in various ways. What motivated me to write this book was wondering why we seem so willing to go along with this system. What surprised me was that the answer goes far beyond convenience, or a fatalistic sense that it's impossible to escape the all-powerful algorithms.

Because we know that we're observed, the technology through which we live so much of our social lives is not neutral or transparent. We treat the platforms very much like that eavesdropper at the table: a bit creepy perhaps but tolerated as a fact of life – especially when they always seem to buy the drinks. We care what they think of us.

'Facebook is offering me slippers,' I moan to my friends,

depressed that the algorithms see me as an old-fogey stay-at-home. I miss the days when all the adverts were hangover cures or sexy pants. 'You're lucky,' they reply, 'I'm getting pension plans.' But why do we care what the algorithms think? Because they are now part of our social fabric. We are social animals, and we care how others see us.

'No, no, you're still young,' my friends reassure me, but they would, wouldn't they? The all-seeing algorithm, which collects my late-night browsing habits and logs the hours I wake up and how far I travel on Saturday nights knows the horrible truth: my all-night clubbing days are behind me.

Social recognition has always mattered to us. That's why we have awards ceremonies. It's why people who have earned a PhD like to be addressed as 'Dr'. But as well as being recognised for things that our society values – virtues or achievements that everyone agrees are worthy of respect – we crave recognition of who we are as an individual. We want to be known for who we feel ourselves to be.

I don't know whether this urge is stronger than it's ever been, but I do know that it's more important in society than it's ever been. By the end of this book I hope I will have convinced you that this is true, and why it matters so much. Never before has the importance of recognition been so central to laws, to the physical environment, and to our idea of what a good society would be like.

This is why we like it when adverts recognise something about us, especially if it's a positive quality. We know that algorithms aren't people, but they are part of our social fabric. They observe us with dispassionate, data-gathering eyes, noting our every deed, word and possibly thought, and see us – supposedly – for who we really are.

It's often assumed that consumers don't like being profiled and targeted, and that 'we offer consumers a better experience' is just an excuse from companies who want to stalk us and sell us more stuff. The reality is more complicated, and more contradictory. There's evidence that we do, genuinely, like being targeted with adverts, and not just because we waste less time scrolling past adverts that are blatantly, insultingly wrong. We like the feeling that the algorithms have recognised in us the positive traits that we value in ourselves.

When I say 'evidence', I don't simply mean that we keep buying things, and the companies behind Google and Facebook (both glorified ad agencies) keep getting richer. I mean that researchers took human beings into a laboratory and experimented on them by showing them adverts.

In 2016, Christopher A. Summers and his colleagues decided to test their theory[27] about the difference that 'behavioral targeting' makes to online adverts. That is, targeting by algorithm, based on that individual's previous behaviour online. Like all psychology experiments, this involved various sneaky manoeuvres, asking participants to answer questionnaires about themselves, select products and travel options from the internet, rate adverts and say how likely they would be to buy the thing being advertised. Research subjects were told that they would see adverts selected either at random, by demographic profiling (age, gender, and so on), or in response to their supposedly free internet browsing in the previous study.

In reality, all three groups saw the same advert (I told you they were sneaky). They all saw an advert for a fictitious restaurant called Eatery 21, offering 'Refreshingly Sophisticated American Classics'. Just as the researchers predicted, people responded differently to the same advert depending on why they thought they were seeing it. People who thought the adverts were demographically targeted were no more likely to say they would buy the product than people who thought the advert was random; a tiny bit *less* likely, in fact. People who thought the advert was being shown to them because of their own individual behaviour in the earlier session were distinctly *more* likely to say they would buy the product.

Far from hating adverts that target us on the basis of our browsing history, we like it so much that we are more likely to buy the product. The researchers explain this as the effect of an implied social label. If we think that an advert has profiled us as 'sophisticated', or 'outdoorsy', or 'environmentally aware' because of our browsing history, we accept that quality as an important part of who we are. Not only are we more likely to buy things that correspond to that quality, we're also more inclined to donate to relevant charities later.

Before you get too paranoid about our vulnerability to online manipulation, this effect only worked when the social label was plausible. An indoorsy person could not be converted to a taste for the outdoors by being shown an outdoorsy ad for hot chocolate – not even when told (falsely) that it was based on their previous online shopping behaviour. It takes more than some sneaky researchers in a psychology lab to lure the committed sofa-dweller out into the muddy cold.

We humans are good at imbuing a product with social meaning. Confused by a sign in an infant school reminding children that *We do not talk about our shoes*, I was told that even five-year-olds knew which brands of trainers were desirable and which were low-status, and that had become a source of social friction. Perhaps the Aterian people kept a sharp eye on the latest cave paintings to check whether their snail shell beads marked them out as trendsetters or made the other men on the antelope-hunting party snigger behind their backs.

But the power of an implied social label from personalised advertising goes one stage further. It invokes the oracular power of an algorithm that observes your most private actions with its impartial, all-seeing, digital eye. It builds on the myth that Big Data knows you better than you know yourself. Giving some positive quality like sophistication the external validation of a targeted advert made people feel more strongly that they were sophisticated, and that had a lasting effect on their willingness to buy the sort of thing that a sophisticated person would buy. In another experiment, the implied social label of being environmentally conscious also made those participants more likely to donate money to an environmental charity after the study was (supposedly) finished.

This contradicts the idea that we don't like algorithms to know us too well. When an algorithm recognises our finer qualities, we like it very much. It changes our own perception of ourselves in a direction that makes us feel good. I suspect that the supposed objectivity of an algorithm makes that effect even stronger, though this experiment didn't test that.

It also reveals that our understanding of how profiling works is more sophisticated than often assumed. When we are told that an advert is targeted at people broadly like us – the same age, gender, whatever – we don't feel at all valued, understood or recognised. We know that we're being lumped in with a large population. We may even resent the assumption that we're interchangeable with everyone else, as if everyone in our class at school would think the same way and like the same things. It's not exactly flattering.

Somebody who's told that an advert has been selected – even by a machine – because of their individual behaviour is getting a very different message. However crude and downright wrong the algorithms may be, they are unswayed by what we *claim* to think, like and be. They take only the evidence of our actions and, from that, see into our souls. So what if your friends think you're a country bumpkin who doesn't know a fish knife from a pastry fork? The algorithm has inferred that you have sophisticated tastes.

As I finish writing this chapter, Instagram shows me a video from a band called the Last Dinner Party. 'If this is on your feed,' reads the caption, 'congratulations that Instagram thinks you have excellent taste in new bands.' How perceptive. I *do* have excellent taste in new bands. Thanks, algorithm, for recognising that.

Chapter 2: From Mass Media to Your Own Channel

'That's one small step for a man, one giant leap for mankind.'

NEIL ARMSTRONG, 1969

Half a billion people around the world watched Neil Armstrong set the first human foot on the Moon. Nearly one in seven of the people alive at the time were briefly united in watching indistinct black-and-white moving images, relayed in real time from millions of miles away, aware that they were witnessing something very important, even if they were too young to understand how much of a first it was for humanity.

As *The Times* put it at the time, 'July 20, 1969, will be remembered when little children who were brought down half asleep are grandparents. It is the first event of such historic significance to be shared so widely and known so immediately.'

Until the late twentieth century, the mass media broadcast through just a few channels – first, newspapers, later, radio and then television – to mass audiences. Once a person had settled on a particular newspaper or a channel, they got the same content as millions of others, in the same order, at more or less the same time. A handful of people decided what merited airtime or column inches, and how those stories should be told.

Of course, we still share media experiences. Over half a billion people have watched a video of a baby biting his brother's finger, filmed on a smartphone. *Charlie Bit My Finger* was the most-viewed YouTube video for several months, not because any editor decided it was as important as landing on the Moon, but because those who saw it shared it, passing it on peer to peer, until it entered common cultural language and became a meme.

Music videos, with marketing departments behind them, do better still. 'Baby Shark' has over 12 billion views, more than the entire population of the Earth. It went viral, spreading through the bloodstream of our digital culture, one share button at a time. That's 12 billion tiny steps for a thumb, one giant leap for an annoying song about sharks.

But we experience today's shared media events in a very different way. Every individual has a *personalised* news channel, edited by their own choices, their social network, and algorithms that predict what they want to see next. Most people under thirty-five get their news through social media and the internet. So, as many commentators have noted, we can no longer presume a shared worldview, or even a shared reality. Each of us relates to a unique world, filtered through technology.

We've also been transformed from passive consumers of other people's stories, images and ideas, to producing and sharing our own content. We are not just an audience, but the writers, directors and stars of our own personal media channel. Today, galleries mount exhibitions, and cities commission public art, designed primarily to make photogenic backgrounds for our selfies.

We have moved from the century of mass media to a world of social media. Not only do we each receive a tailored, digitally mediated view of the world, but we increasingly live our own lives within a world structured by technology. Within that world we interact with news and entertainment, but we also interact with each other in new ways, as each exchange helps create our persona, an online identity that expresses who we are and how we want to be seen.

*

Broadcasting House, the BBC's home since 1932, has weathered direct hits from wartime bombs. One ripped out a section of the fifth floor, killing seven people. Newsreader Bruce Belfrage paused briefly as he heard the explosion, then continued the nine o'clock bulletin without reacting.[28] Through military and political conflict, social and technological revolutions, broadcasting has continued from this spot. But on this cold day in 2022, as the BBC celebrates a century of Public Service Broadcasting, I am wondering whether that mission has a future in the twenty-first century.

I meet Professor Jean Seaton, the official historian of the BBC, in the little courtyard café. On the table I put our coffees and my phone, to record our conversation. I never did learn shorthand, because by the time I started working as a journalist, it was normal to record everything and transcribe later. At first, that was on a little machine using micro-cassettes – little cassette tapes about the size of a USB stick – then minidiscs, digital recorders, and finally, my smartphone. Some of my adventures in audio technology have been making radio for the BBC.

I ask Professor Seaton what marks the shift from mass media to today's more personalised experience, expecting her to talk about the on-demand iPlayer or the BBC Sounds app, or online news. Her answer surprises me: 'The Walkman.'

For anyone who grew up in the twenty-first century, to whom that word means nothing, a quick history lesson: in 1977, a German-Brazilian inventor called Andreas Pavel filed patents for his invention, the Stereobelt, a 'stereophonic production system for personal wear' consisting of a belt, battery packs, a playback device, amplifier and headphones. He failed to sell the idea to a big technology company, later telling the *New York Times* that 'they didn't think people would be so crazy as to run around with headphones'.[29]

But two years later, Japanese company Sony released the Walkman, adapted from their Pressman portable cassette recorder for journalists. The first Walkman simply replaced the record function with a double headphone socket – so you could listen with your friend – and a microphone to let you talk to each other over the music. Later models dropped the microphone altogether, finding that most people listened alone.

As someone who got a Walkman (or a cheaper imitation) at some point in the 1980s, I can tell you that it revolutionised my relationship with sound. Instead of having to go where the sound was – a shared record player or radio, or a live performance – I could add the soundtrack of my choice to whatever I was doing. As Seaton says, 'Instead of listening to music at home, you could put a cassette on your Walkman and walk around with it.'

Just walking to the shops suddenly felt like being in the movie of my life. I could enhance, or try to shift, whatever emotions were rampaging through my teenage brain. Instead of the boring, everyday sounds of the people and places around me, I could live inside a heightened version of my own private experience. Today, it seems completely normal to move through the world with your own personal soundtrack, whether that's music, podcasts, a phone conversation, or the audio for whatever you're watching on the little screen. Perhaps you're even listening to this as an audiobook while you do whatever else it is that you're doing.

It seems astonishing, looking back from today's world, that tech companies didn't think there would be a market for personal soundtracks. The success of the Walkman took even Sony by surprise. Japanese stores sold out within a month of first release. Within ten years, they sold fifty million units worldwide. As well as compact, durable technology, Sony had clearly picked up a change in what people wanted; what Seaton calls 'a sociological re-understanding of individuals'.

'You could say that the Walkman is the result of the sixties,' she suggests. 'The sixties had within them a sense of individual rights against collective rights.' But not for poor Andreas Pavel, who pursued Sony in court for breaching his patent rights, with mixed success.

We've never had a completely homogenous media experience. We might have relied on a handful of newspapers in the hands of a few people for our hard news, but for entertainment and niche information, there was a choice. From the early twentieth century, magazines and newspapers have catered both for mass audiences and for specialist interests. Photography, fashion, motor cars, movies, politics, literature – whatever your interests, you could find a publication to entertain and inform you. There was also local provision, so you could get the news about yourself and your neighbours.

In the United States, commercial television stations became active in the 1930s, licensed by the Federal Government, and broadcast to local audiences within reach of their transmitters.

The American airwaves became so crowded that the issue of new licences was frozen in 1948, and local cable stations sprang up to relay existing channels to new audiences. Much later, the proliferation of British television channels, from three in 1980 to hundreds via cable or satellite, gave UK television audiences a similar degree of consumer choice.

The advent of cable TV was, prematurely, predicted to transform British social life. In a 1982 speech, then Prime Minister Margaret Thatcher foresaw a range of two-way services: 'It's possible to do mail order by cable; to read a train timetable by cable; even to choose your holiday – complete with colour pictures of the surf breaking outside your hotel.'[30] Possible, perhaps, but not that appealing to most UK viewers. Hardly anyone was willing to pay for extra television channels when they could watch free broadcasts. And for rolling news, weather, TV listings and flight arrival times they could turn to Ceefax and other teletext services on their television screen. Started in 1974, Ceefax was originally intended to provide subtitles for BBC television programmes,[31] and its twenty-four pages had to be loaded from punched tape, but it quickly found a wide and loyal audience. Its demise in 2012 was met with nostalgia for the clunky graphics and slow rotation of pages.

Cable television's UK successor, satellite television, got off to a slow start in the 1990s, gradually building audiences with a diet of news, sport, cartoons and music. The move from analogue to digital television was also a reluctant revolution, with viewers slow to spend money on new equipment. Eventually, a combination of free digital channels and the experience of the internet wooed the British public to the new television world. The pandemic boosted our tendency to watch a wider range of onscreen content, but already in 2017, nearly half the adults in Britain had a subscription to a television streaming service. Today, we choose between channels, on-demand streaming services and apps, and within them choose not only what to watch, hear or read, but where and when to start and finish the experience.

Netflix, originally a mail-order service for renting DVDs,[32] is an on-demand service with a recommender system. In fact, Netflix was the first company to make many people aware of how recommender systems work, thanks to the Netflix Prize. With a finite number of DVDs to mail out, it was important to keep customers satisfied with the films they did receive, while they waited for their first choice. That meant predicting which films would satisfy which customers. In 2006, Netflix announced a $1 million prize for any team that could improve its existing Cinematch program, designed to predict how any given customer would rate a given film, by 10 per cent. By the time the final winners were announced in 2009,[33] however, Netflix was no longer interested in improving its existing recommender system, because in 2007 it started streaming content.

'Streaming has not only changed the way our members interact with the service, but also the type of data available to use in our algorithms,' explained a Netflix Technology Blog in 2012.[34] 'For streaming, members are looking for something great to watch right now; they can sample a few videos before settling on one, they can consume several in one session, and we can observe view-ing statistics such as whether a video was watched fully or only partially.' Thanks to this instant feedback, Netflix rapidly achieved a personalised service from the moment you start wondering what to watch. 'Personalization starts on our homepage, which consists of groups of videos arranged in horizontal rows . . . Each row represents three layers of personalization: the choice of genre itself, the subset of titles selected within that genre, and the ranking of those titles.'[35]

People – including me – often complain that algorithms are black boxes, making nefarious decisions that affect our lives, with no explanation or appeal. Netflix, by contrast, makes a point of telling customers why they have selected films for recommendation – what you told Netflix explicitly about your tastes, what you've watched before and – if you have allowed Netflix to connect with your friends, on Facebook perhaps – what your friends recommend.

Most of what I watch on Netflix is precisely what my friends recommend, but they've recommended it directly to me when I wasn't near a television. When it comes to letting the recommender systems choose my media content, a better example for me is Spotify. I've been delighted with its ability to extrapolate from a few artists or even tracks that I liked, and feed me new music that I had never heard before, but am glad I did.

Having remembered that I like Fela Kuti's music, I asked the algorithm for more like him, and now have a rich and varied audio diet of Afrobeat, spiced with wonderful oddities like Lord Kitchener's 'Come Back With Me Wife Nighty'. If I like a particular track, I can tell Spotify so, or add it to a playlist for future enjoyment. As far as I'm concerned, this is a glorious example of a service tailored to my personal tastes with no downside, except perhaps that the artists get so little money from my listening that they may stop making new music.

Anthropologist Nick Seaver has been looking at recommender systems almost as long as Netflix and Spotify have been using them, though the principles go back further still. 'The premise of "we're going to find people who resemble you in some way, based on some kind of data that we're collecting, and we're going to try to show you the things that those people seem to like," that's been there for almost thirty years now,' he tells me.

Nick's book, *Computing Taste*, is a thoughtful exploration, as well as a history, of music recommendation systems in particular, and the people who create and apply them. He is sceptical of the idea that a computer can capture a person's taste in music. In fact, he's sceptical that there is such a thing as a person's authentic musical taste outside of the social world they live in, what he calls 'some pure taste in music that's unshaped by the designs of others. The music industry, your peers, what music gets made, everything is shaping you all the time.'

In other words, there is no 'what you like' for the predictive algorithm to capture. From day to day, even from hour to hour, how you feel about different music is changing. And the people who make recommender systems are getting wise to this, says Nick. 'I think it's much more common for people to say, maybe your preferences vary contextually.' Spotify doesn't have one all-powerful algorithm to recommend music to listeners. It uses a combination of different computer programs and human experts in different types of music. Unlike social media platforms that rely on user-generated content, somebody has selected, or at least approved, each track before anyone gets to stream it.

Nevertheless, the way recommender systems understand music is never the same as the way listeners understand music. Because of this, Spotify's role in how music is distributed and heard is causing subtle but potentially important changes. Nick gives, as an example, 'what it means to be post genre in the current moment. We're genre fluid now, anything goes.' But in this post-genre music, Nick sees 'a wild flourishing of genre terms. So it's this weird paradox, where on the one hand, you're saying labels don't matter anymore. And on the other, you're getting so many labels, and at the same time you're getting a real change in what musical genre means.'

As somebody who both makes things for audiences and likes to go to the cinema, I have a very practical idea of what genre is, and what it's for. I think of it as a relationship between the audience and the person or team making the film, or radio programme, or whatever. If I'm told something is a romcom, I have certain expectations before I sit down in the cinema. By the time the credits roll, I want the two protagonists to have got together, and for neither of them to have died in a shootout, or any other way. Apart from that, I'd like you to surprise me.

Genre is part of a language that the creator uses to communicate with me. Genre is part of the reason why a movie, or a song, is like it is. Directors and musicians can subvert the genre, or play with it, giving me a Western that leaves me questioning the morality of shooting people, or a love song that insists the singer's not in love, but that's a deliberate creative choice.

'But now,' says Nick, 'it's possible you put music on Spotify, and learn from Spotify that you are in a genre you did not know you were in.' Your cover of a Talking Heads song pops up in a playlist called 'Folk Fabrique' and now you are hanging out, digitally, with Caribbean Jazz and Tuareg drummers. That's not your choice, it's Spotify's choice. 'That changes what genre means to become this post hoc thing, where it's a consequence, rather than a cause, of similarity. So to put it mildly, that's a very different definition of genre than one that people have historically used.'

I suggested that Nick was sceptical of the idea of a person's authentic, pure taste. But in fact, he goes much further than that. Without our world of consumer choice in music, he's sceptical that such a thing as taste could even exist. 'Imagine that it's 200 years ago, and you have taste in music. What does that mean? What are you doing with your music taste? There's no such thing as recorded music. You've got people who play music around you, you might play music, you might have preferences for songs, but you can't really have taste in music, the way we think about it today.'

Taste, for Nick, helps you choose between interchangeable products: 'Imagine you go to a record store, here's a bunch of records, they're all the same shape. They all work the same way. There's no reason to buy one of them over another one, except for this sense of taste.' It's a new way to express myself by choosing one record – or CD, or Spotify track – over another. And that choice will further shape my ever-changing taste in music.

What I think of as *my* musical taste is a constant interaction between the choices I make, and the social and cultural context in which I live. And, as Nick points out, that context includes Spotify, which, at the moment, is only part of my musical environment and influence, along with the radio, other people's music and (let's be brutally honest here) anyone I'm currently trying to impress by looking cool and musically discerning.

But as Nick says, we're approaching a world where recommender systems are everywhere. What happens when they become so much more than a novel tool to engage with our own discoveries from the outside world? I brought Fela Kuti to Spotify from my non-Spotify life. But what happens if I no longer have a non-Spotify musical life? The recommender system becomes one giant mirror, a hall of mirrors, reflecting itself with me in the middle.

I'm playing Spotify as I write this, specifically a Daily Mix that the algorithm has put together for me. It's just chosen to play me Gil Scott Heron's classic, 'The Revolution Will Not Be Televised'.

'The revolution will not be brought to you by Xerox in four parts without commercial interruptions . . . '

Spoken, rather than sung, over a funky, jazzy instrumental underscoring, he lists all the television and movie tropes that will not happen, all the advertising claims that the revolution will not meet, and the things that viewers will not care about:

'The revolution will put you in the driver's seat
The revolution will not be televised, will not be televised
Will not be televised, will not be televised
The revolution will be no re-run brothers
The revolution will be live.'

I click the heart, to tell Spotify that I like this track, and I wonder what, in my data, made the algorithm choose to play it to me now.

*

In the 1930s, newspapers used market research surveys to find out what their readers liked, and the answer was that human interest stories – crime, personal tragedies and scandals – and sport attracted far more readers than what they called 'public affairs'.[36] Now, as then, we turn to media for many reasons, and to meet many needs. Being entertained, moved, reassured, stirred to outrage, connected with others or given reasons to feel superior to others may push us to pick up a newspaper or, today, a smartphone. Being informed is only a part of what we want from the media, in whatever form.

The media have always needed to attract an audience to fund their existence, whether through sales, subscriptions or advertising. Netflix's announcement of a cheaper subscription model if you're prepared to watch some adverts is a reminder that we're often more profitable as a captive audience than as a paying customer.

Media dependence on advertising goes back as far as the nineteenth century. Ever since large numbers of people have had money to spend, and a choice about where to spend it, advertisers have competed for our attention. British advertising spend almost tripled in the thirty years between 1908 and 1938,[37] and most of it went to newspapers. In the US, where commercial radio and television were already reaching significant audiences in the 1920s, advertising spend increased tenfold in the same period to almost $3 million.[38]

'Historically,' says Jean Seaton, 'advertising, right the way back to the end of the nineteenth century, certainly through the twentieth century, on steroids by the fifties, sixties, seventies, eighties and nineties, attempted in various ways to pitch goods and services at distinctly identified groups of people.'

This means that size of audience was not the only way to be commercially successful. Getting the right audience, ideally those who were not already being monopolised by another channel or publication, could also work. Seaton makes this point about Channel 4. The UK's fourth TV channel was founded in 1982 with the mission of diversity in content and audience, reaching minority groups not well served by existing television channels.

Which, at first glance, may not look like the most lucrative advertising market. People from ethnic-minority communities, and disabled people, tend to be under-represented in groups with high disposable income. But their specific needs and interests also tend to be under-represented in mass-market advertising. Research by Channel 4 in 2020 still found that 'Two thirds of BAME people say they'd feel more positive about any brand that showcases different cultures in their advertising.'[39] So, though minority groups, by their nature, may be smaller in number, well-targeted advertising stands a good chance of getting results.

As Seaton points out, Channel 4 would also appeal to anyone who valued more diverse content and wanted to be watching a channel that reflected their social values: 'If you look at the foundation of Channel 4, the big argument around advertising was that you could sell Channel 4 to a group of metropolitan, cosmopolitan, middle-class people who wanted quite valuable things.' ITV was addressing a mass audience, but by appealing to niche audiences, some of whom had more money to spend than ITV's viewers, Channel 4 could be viable as a business model on income from advertising.

'Advertising has always sought to find and identify communities to whom you could sell things, and then enhance the image of what you're selling, to fit into them,' says Seaton. 'In one way, viral advertising driven by algorithms is only a steroid-enhanced version of that.' Advertising spend in the UK hit £30 billion in 2021.[40] Worldwide, companies spent over $700 billion on advertising in 2021, over half of it digital.[41] The biggest tech companies, which dominate the highest-valued companies in the world, get most of their revenue from advertising. The place where we go to see much of that advertising is not just the internet, but social media.

As soon as the internet got going, tech-savvy enthusiasts were using it to post, make new virtual friends, and interact with them as much as the technology would allow, but it was in the early 2000s that sites like Friendster, LinkedIn and MySpace made the online social experience available to ordinary people who happened to have a home computer.

In 2006, MySpace was the most visited website in the world, but it was about to be usurped by another site, founded by a few Harvard students two years before. Facebook, originally open only to students in exclusive American universities, was opened to anyone over the age of thirteen in 2006, and overtook MySpace's visitor count in 2008. The advent of the smartphone made it easy to check in at any odd moment, anywhere. In 2022 almost 2 billion people logged in to Facebook every day.[42]

The pandemic accelerated the trend towards interacting online. In 2021, Americans spent an average of 4.2 hours per day on their mobile devices, only slightly ahead of Brits with four hours. Brazil and Indonesia topped the table with an average five hours a day on their mobile phones. Of that time, 70 per cent was spent in social or video apps. The average American or Brit was spending almost three hours a day on social media.[43]

Now, it's Facebook's turn to face competition from TikTok, the video-based site launched in 2017, which was the most downloaded app in the world in 2021, in spite of being banned in India. The average TikTok user spent nearly twenty hours a month on the app in 2021, the same amount of time as the average Facebook user, and longer than any of the other top-five apps.[44] Two thirds of American teenagers said they used TikTok in 2021, but only a third were still using Facebook.[45]

A 2022 survey asked people how they first came across news in the morning. One third of people in the UK and Ireland, and almost as many in the US, said 'smartphone'. Among under thirty-fives, almost half in all countries surveyed turned first to their smartphone,[46] where the most popular news source was social media, especially Facebook and video channels like YouTube and TikTok.[47]

This is the direction of travel, and a challenge for what are sometimes called 'legacy media' organisations like the BBC or newspapers. Just getting content online, even onto phones, is not enough. The generations coming of age in the personalised century will not go to a news website, or open a broadcaster's app, to get just one version of the news.

This media experience of continuous, fragmented material from myriad sources, with no editor except social networks and algorithms, is a significant change from a century ago, not just for the recipient of news and entertainment, but for the organisations producing and distributing it. Newspaper editors never knew exactly who read which articles, let alone who read to the end or passed the paper across the table to say, 'Have you seen this?' Radio programme makers didn't know (and, for broadcast radio, still don't) who listened from start to finish, where they listened, or who else was listening with them.

Today's digital platforms provide a two-way flow of information. Video apps and channels know how long people spend with a video and build up a record of which other videos the same person watches. News websites can tell how long you spend on an article, whether you scroll to the end, and who you send it on to (and by which routes). Podcast apps know not only what your favourite podcasts are, but where and when you listen to them.

For content producers, this is measurable feedback. For advertisers, it's valuable data. Today's social media adverts, promoted posts and influencers are part of a long tradition and a powerful industry, but in a new form with radically new capabilities.

As Jean Seaton reminds me, 'The scale is new; the invisibility is new. The algorithms matter, in the sense that they drive you towards more of what you like, whereas in the old media environment, and the old political environment, you were, willingly or not, obliged to encounter alternative points of view. It is the algorithms that give you – in order for somebody else to make money – more of what you like. You see it in journalism clickbait. It absolutely drives the commercial model.'

I share Professor Seaton's concern about the opaque shaping of our individual media channels for purposes that are not our own. But there is cause for optimism about the scale and nature of its impact.

I first met Professor Chris Bail at a conference in Silicon Valley, where he was speaking about the work of his Polarization Lab at Duke University. He and his colleagues were, as the name suggests, researching how social media feeds political polarisation. That was in February 2020. Since then, the impact of the pandemic gave them a laboratory in which to study the behaviour of Americans who were suddenly deprived of in-person contact and spending more time online.

The Polarization Lab found that Americans, when surveyed, showed a fairly united front against Coronavirus. Well over four in five Americans supported anti-pandemic measures, including closing borders and restricting social contact, almost regardless of which political party they supported. Only one in five still thought their country was 'very divided' in April 2020, down from three in five just two years before.[48]

Over on social media, however, extreme political positions and conspiracy theories were dominating public exchanges, while moderate voices were quiet. Entirely predictably, the pandemic and subsequent social distancing accelerated existing trends towards online as our main social space, for better and worse. This has fed calls for more regulation of potentially harmful content by tech companies, and of tech companies by governments.

Professor Bail's 2021 book, *Breaking the Social Media Prism*, should be on the reading list for anyone trying to draft such regulation. His Lab has done in-depth research with individual social media users from across the political spectrum, as well as collecting statistical evidence. He concludes that 'there is surprisingly little evidence to support the idea that algorithms facilitate radicalization'.[49]

Only 1 in 100,000 YouTube users move from viewing moderate to extreme material. Less than 1 per cent of Twitter users were exposed to most of the 'fake news' shared during the 2016 US election campaign. Interacting with Russia-linked social media accounts had no significant effects on political attitudes or behaviours.[50] What online interaction *does* feed is what Bail calls 'false polarization' – the perception that the other side's views are further away from you than in fact they are. 'Democrats and Republicans also overestimate how much people from the other party dislike them,'[51] which naturally makes those other people more dislikeable.

The idea of the echo chamber or filter bubble, the narrow window on the world that reinforces the viewer's existing world view, leapt to Professor Bail's mind in November 2016: 'The day after Trump's improbable victory . . . my mind kept returning to one simple fact: how had so many of us never seen a shred of evidence that Trump could really win? What were we not seeing that made us so surprised by Clinton's defeat? The concept of the echo chamber provided an elegant explanation.'[52]

It's a stark reminder that no social media network reflects the full spread of views in a country's population. Twitter users, for example, tend to be young, educated, affluent and liberal, relative to the whole population, though the same may not be true now it's become X under new ownership. But in reality, few people live in a bubble of homogenous views, even online.

A month before Chris Bail's shock discovery that millions of Americans voted for Donald Trump, a survey found that most Americans' Facebook and Twitter networks include a mix of political beliefs. Though two thirds reported that discussing politics on social media with people who disagree is stressful and frustrating, one third said it was interesting and informative. Disappointingly, the more politically engaged social media users were also more likely to have blocked, unfriended or muted somebody because of politics.[53]

A UK study two years later found that only 8 per cent of participants could be said to live in an echo chamber online, in the sense of rarely encountering things that cause them to change their minds.[54] Another large US study concluded that finding news and opinion pieces via social media or search engines led to *more* exposure to opposing points of view, not less.[55] Compared to previous generations who wouldn't dream of picking up a newspaper with a different political perspective, we are in some ways less insulated from the diversity of opinion in our society.

Chris Bail and his team were surprised by many of their own research findings. He has not stopped being concerned about the effects of a fragmented media and social media environment, but he has come to some new conclusions about what we should be doing about the problem. I'll return to those in a later chapter.

For advertisers, and the platforms that make their money from advertisers, social media has an advantage over older forms. We find it much harder to tear ourselves away from things that have been written, photographed or shared by people we know. Social media runs on user-generated content: that is, what you and I post, and what our friends, family and strangers post.

Teenagers, when asked what they value about social media, tend to talk about connection.[56] It's easy to forget, when looking at social media platforms as business models based on getting and keeping our attention to show us advertising, that they work so well because they meet a very basic human need. We are social creatures, and social media platforms enable us to interact, socially, with other humans. Another powerful force that draws us back to social media is the desire to get a response to what we posted ourselves. We are not just consumers and editors; we are the stars of our own channel.

The more we conduct our social lives through screens, the less we can distinguish our personal lives from our public posts. This imbues our online interactions with emotional intensity that they shouldn't always have. Do I really care whether a complete stranger thinks my silly pun isn't funny? I shouldn't, but I do. Like it or not, we understand ourselves in relation to our social world. Today, that world is likely to be at least partly digital, which both changes and reflects how we see ourselves in the twenty-first century.

As Jean Seaton and I finish up our coffees, I ask her about the future of public service broadcasting, when all three of those elements – the public, service and broadcasting – have completely changed in the last hundred years.

'I think service is a fantastically valuable thing,' she replies. 'It doesn't mean just Ocado giving you your food, it means the honourable, absolutely life-enhancing, personal and institutional devotion to something above you, which is serving the public.' This, perhaps, is why she is also director of the Orwell Foundation, a charity set up to 'perpetuate the achievements of the British writer George Orwell',[57] author of *Animal Farm* and *1984*.

However, when we move on to defining 'the public', things get trickier. There are, says Seaton, 'lots of publics' but there are also some things that are best provided universally. 'There's a wonderful Fabian essay from about 1910,' she tells me. 'Beatrice Webb and Leonard Woolf agree that they've watched milk carts go down their roads in Bloomsbury, and they think how ridiculous it is. Because there are many milk carts, and they're competing with each other. And wouldn't it be more sensible if there was one milk cart?'

That is exactly how I feel about delivering letters and parcels. But I'm not sure that a milk cart is the best analogy for the media. Reliable news is a vital part of public life, not just a product for individuals to consume, but that doesn't mean that we need just one news channel. Having access to diverse sources of news lets us compare different ways of understanding our shared world. The one milk cart for news would leave only one milkman to decide what does or doesn't count as fresh milk, or milk at all. What if he decides that oat milk is Fake Milk, and refuses to deliver it?

On broadcasting, though, I agree with Seaton that there are still times when we all come together around a real-time event. The funeral of Queen Elizabeth II is one example, the World Cup another. The technological medium is less important than the shared experience.

She ends our conversation with some very personal reflections on the capacity of individuals to see what's right and to stand up for it, instead of allowing themselves to become complicit in evil regimes. 'This kind of moral individualism is unbelievably valuable, and you see it around journalists – there are people who get killed for what they'll do,' she says. 'But most of us aren't heroes. I'm not a hero. I've got no moral fibre at all. Orwell had no complicity, there was nothing complicit in him. He was an awkward bugger. The rest of us are just fallible. And so, I think this individualisation, yes, there are valuable things you can do with it. But I wouldn't overemphasise our capacity to think freely and individually.'

On this depressing note we part, Professor Seaton back to the University of Westminster, and me to my homeward train. I look around at my fellow train passengers. Most of them are absorbed in their phones, some reading books or newspapers. We may all be sharing a public space, physically, but our minds are in myriad places. Do I have faith in their ability to think freely and individually? Yes, I think I do. But I wish we could bring our free, individual thinking to more public discussions that aren't social media shouting matches.

*

On the BBC's hundredth birthday, as it happens, I am working in Broadcasting House, making a radio programme. I'm disappointed to find that there is no in-house celebration at all, not so much as a piece of cake to take home, or a balloon. I know the BBC is strapped for cash, but I can't help feeling it reveals a lack of self-belief.

I go for lunch with Bill Thompson, who is a mainstay of BBC Research and Development, and a lifelong technology innovator. For Bill, technology especially means the internet, in which he's long been a key player. 'I'm the person who made the *Guardian* free online,' he tells me. 'I'm one of the people who was behind the [BBC] iPlayer, originally.' Now, he's working on the next generation of innovations that will change our media landscape.

In a nearby café, over meatballs and mash, I ask him what he thinks the next century holds for the BBC. The new technological environment is a challenge to an organisation that spent a century getting very good at making radio and then television programmes. They no longer have a guaranteed place in the lives of anyone in Britain with a television set. 'If you turn on your Sky Glass,' says Bill, 'it'll give you the user interface, which foregrounds the programmes it has decided you might want to watch, and some of them might be BBC programmes. But getting to iPlayer and then getting to the BBC schedule is several clicks. And that's a real barrier.'

I ask him the same question about the future of public service broadcasting that I asked Jean Seaton. Bill thinks the BBC's understanding of the public 'has absolutely become much more sophisticated and practical, particularly trying to encompass the diversity of opinions and backgrounds, everything like that,' but he's also keen to point out that the BBC never did see the British public as homogenous. There was always a concern to reach working-class people, in particular, and to grapple with what 'universality' meant in practice. Nor was the BBC of old a top-down disseminator of the great and the good with no voice for anyone else. 'The BBC used to get hundreds of letters a week in the [nineteen-]twenties,' Bill reminds me. 'There was a whole department dedicated to dealing with that correspondence coming in.' Now, that feedback via social media is constant, instant and public.

But Bill's idea of service is no longer about putting entertaining and educational programmes in front of an audience, like serving plates of food to hungry people. In his vision, the internet is more than just a new table, where diners can get a twenty-four-hour buffet instead of consuming a set menu at set mealtimes. He thinks service means a lot more than 'the idea of a platform as being a place where you put it to serve them'.

One of his main concerns is how to make the internet a better place for many voices to be heard, and how the BBC can play a role in making that happen. 'We can serve the public by helping shape the online environment,' he says. 'What are the other things, what additional value can we bring to people's lives using the resources, the editorial and engineering capability of this amazing organisation, to do things that matter? That sustain a digital public sphere, where civil society can flourish, and we can have the multifarious debates from all of our different traditions and reflect the complexities of society.'

Some kind of public service internet, then?

'Yes, precisely,' says Bill. 'And where you can talk to each other and express your creative intentions and feel safer than with the current internet.'

Providing a space for public conversations is almost the polar opposite of the old public service broadcasting model, which Bill describes as the idea that 'we'll provide material, which is the starting point for discussions and debates between you all, in spaces we know nothing about and have no access to – your homes, your workplaces, your political meetings. We didn't actually know who was watching. We didn't know what they were taking away from it. We were just doing our best, throwing it into the darkness from our sticks on the hills.'

As Bill and I walk back to work, we pass a statue at the corner of New Broadcasting House. It's a bronze figure, a man in a crumpled suit leaning forward from a high plinth, cigarette in hand. He could be about to walk away, or just a little restless on his feet.

George Orwell worked for the BBC from 1941 to 1943, broadcasting to the Indian subcontinent where he was born, the son of a colonial administrator. In wartime, he was essentially making propaganda for the British Government.

Orwell had already experienced war as a combatant, having fought in Spain with the International Brigades against General Franco's right-wing Nationalists. He also had to flee the communist factions on his own side who, backed by Stalinist Russia, were suppressing other left-wing groups and voices. No wonder he had a lifelong wariness of governments telling people what to do and say and think.[58]

There's usually a human smoker or two nearby, looking down at their phones, not up at the inscription on the wall beside Orwell.

If liberty means anything at all, it means the right to tell people what they do not want to hear.

Chapter 3: From Mass Movements to the Personal Is Political

'The 2016 election cycle will be remembered for many things, but for those who work in politics, it may be best remembered as the year that political data reached maturity.'[59]

ANDREW THERRIAULT

In 2004, Andrew Therriault was a twenty-two-year-old American graduate, and he wanted to help get a Democrat president elected. 'I got in my car, I drove to Ohio, I walked around college campuses, signing students up to vote and all that.' But despite the efforts of Therriault and all the other volunteers who were supporting his campaign, Democrat candidate John Kerry did not get elected.

So Therriault 'wound up after the election, back at my parents' house, broke, depressed and unemployed. And one of the things I took away from that experience was, I liked politics, but I hated talking to strangers about politics. That was not my forte.

'I wouldn't say I'm an introvert,' he told me. 'But that was a very special level of social anxiety for me. The same way I'm not a sales-person, trying to convince people of things just didn't feel right to me. So, I figured, I want to find a way to do politics behind the scenes. What could I do? I've always been good at science and math and computers and that kind of stuff. So how do I use that? I tried to figure that out, and I didn't really know.'

Because he didn't know, Therriault went back to studying for a PhD in political science. His dream job didn't exist yet. 'In 2006, the only data stuff that was happening on campaigns was a little bit of survey research and polling, and some fundraising calculations. Back then we weren't talking about a lot of micro-targeting or experimental testing for ads or anything like that.' But by the time he was *Dr* Andrew Therriault, it did exist. By 2014, Therriault was Director of Data Science for the DNC – the Democratic National Committee, the organising body of the US Democratic Party. By 2016 he was editing *Data and Democracy*,[60] the O'Reilly publication from which the quote above is taken. It describes in detail exactly how data can be used for political campaigning.

With hindsight, the 2016 election cycle was probably not best remembered as 'the year that political data reached maturity', though it is true that, because of people seeking explanations for the election of Donald Trump as president, and the vote for the UK to leave the EU, it was the year that many people discovered for the first time how political data is collected and used. It certainly wasn't the first year that such data was used, though. In 2008, while Andrew Therriault was studying Political Science, Barack Obama was getting himself elected president by combining his desire to mobilise mass support with the new data profiling methods.

By 2012, 'Obama's campaign began the election year confident it knew the name of every one of the 69,456,897 Americans whose votes had put him in the White House,'[61] wrote journalist Sasha Issenberg. This wasn't because they breached the secrecy of the ballot box – not directly, anyway – but because Obama's campaign had enough data on individual voters to predict who was most likely to have voted for him in each district. Issenberg's article spells out how the Obama campaign combined survey information with social media data, how they got around privacy laws by giving the names of target voters to cable TV companies who would then show them particular ads, and much more. Issenberg's article ran in a January 2013 issue of *MIT Technology Review*, with a cover headlined 'Big Data Will Save Politics'.

In the UK, too, the left-leaning *Guardian* newspaper was enthusiastic about Obama's 2012 data-driven campaign, which asked volunteers to share their Facebook credentials. 'Consciously or otherwise, the individual volunteer will be injecting all the information they store publicly on their Facebook page – home location, date of birth, interests and, crucially, network of friends – directly into the central Obama database,'[62] said a gushing article that spared one single paragraph for potential privacy concerns.

Let's go back to Andrew Therriault to tell us from the inside how a data-driven election campaign works. I was talking to him in September 2021, when he didn't seem at all introverted or reluctant to talk politics (though we were talking over Zoom – perhaps that helped).

Micro-targeting in political campaigns, just like the marketing and media recommendation systems in the previous two chapters, works by gathering enough data about somebody to build a model of what they like, what they're like and what they're likely to do. Then the message is tailored to the person who will receive it. In the run-up to a vote, campaigners want to know two things about you: how likely you are to vote their way, and how likely you are to vote at all.

Therriault's job was to build computer models that gave each voter in their files a score from 0 to 100 on each of those measures.

'The part that gets a lot of attention was the machine learning algorithms, and how do you come up with these predictions,' he tells me, 'but honestly, most of the work was on the data itself, which is true of all data science, but particularly in politics.'

I ask him how many data points he would have on a typical voter – dozens? Hundreds? Thousands? 'Thousands,' he says, as if that's obvious. 'Thousands.' But to think of that as a list of thousands of things he'd know about you, individually, would be misleading. Yes, it would include your voter registration details – name, address, age – and perhaps answers you've previously given to surveys. All those can be confidently linked to you, personally. But to that basic voter database, Therriault's team would add census data for the address, which tells him about the neighbourhood in which the voter was living. So, he might know the racial mix of your street, but not your own ethnic background; or the average house price for your postcode, but not what you paid for your home, if you own it at all.

Political campaigns also regularly use commercial data from data brokers like Experian and Acxiom, collected from our interactions with businesses and compiled for marketing and research. 'Honestly, those aren't worth as much as you think,' Therriault tells me. 'It makes for a great story to say, "Magazine subscriptions are really telling." If that's all you have, sure. A common one that comes up is if you subscribe to a hunting magazine, that means you're far more likely to vote Republican.

'The thing is, only a small proportion of people subscribe to hunting magazines. So, it doesn't tell us much about the electorate as a whole. And most of the people who do are already registered Republicans, they live in highly Republican areas, they tend to be older white males who live in rural housing. We don't need the magazine subscription to say that person is voting Republican.'

But those databases are useful for some things: income, marital status and education level, for example. Even though it's hard to match specific individuals from commercial data sources with individual voters, they can be used to build models, so what they *do* know about you can be used to predict things that they *don't* know.

So, let's assume that political campaigners have you in their database, with thousands of data points that may or may not apply to you, personally, but which give them a pretty good picture of who you are. Now what?

Being able to profile individual voters lets a campaigner decide who to target, and tailor the message to them. There are two kinds of people worth targeting: the ones who might be persuaded to support your side, and the ones who need a push to get out and vote. 'Persuade or mobilise,' as Andrew Therriault puts it. For him, knowing what you want a voter to do should govern how you communicate with them. 'If you look at my voting record, I vote in every election. I just cast a vote in the municipal preliminary election in Boston, which is today, which is the lowest of low-turnout elections. Nobody shows up except me. I'm not a mobilisation target, I'm a persuasion target.'

His recent experience, he says, is 'a great example of the individualisation of voter contact. As a high-turnout voter, my mailbox is full of mail from the seven or eight different candidates for mayor and the seventeen candidates for city council. Except for one candidate, a woman named Michelle Wu, who's running for mayor.'

Why is this, I wonder?

'I am a substantial donor to her campaign. And somebody there was smart enough to recognise I have an automated monthly dona-tion. It's like, "He's donating every month, we don't need to persuade him. And we don't need to mobilise him. If he doesn't show up, it's because he's in the hospital." I thought that was a fascinating example of data done right. Because I've done the targeting for those direct mail campaigns; those aren't cheap. So, it was really impressive to me that the campaign I'm donating to understood, "We should save our money and screen out our donors." It shows how nuanced targeting can get.' It's also very effective. Michelle Wu was elected Mayor of Boston in November 2021.

Data-driven campaigning, profiling and micro-targeting are not new, and they're used by parties and campaigns right across the political spectrum. But if they're so sophisticated, and so well stocked with thousands of data points, shouldn't we worry about being manipulated? Knowing my address and that I once voted for you is one thing. Knowing that I am neurotic and introverted is something else. After all, Alexander Nix, CEO of Cambridge Analytica, claimed to have 'profiled the personality of every adult in the United States of America'.[63]

'That was bullshit,' says Therriault. 'I didn't think Cambridge Analytica was going to be the story it became.' He had particular reasons to think it was old news. When the story came out, he had just started work at Facebook, and he knew that Facebook had shut down access to the kind of data Nix was using four years earlier. 'I knew this because I had been using similar data.' Therriault had worked for a survey firm that developed a tool for targeting voters in other countries. 'We did Moldova and Estonia and Ireland, and a few other places where we didn't have voter files to work from. We would scrape Facebook data to get pretty sizeable lists of adults in those countries, and based on things they like on Facebook, you can come up with a pretty good model.'

But in 2014, Facebook changed the terms of data sharing, and that whole project shut down. Companies like Cambridge Analytica would no longer have access to data on our likes and our networks of friends. I ask Therriault why Facebook made it harder for other organisations to scrape users' data, and he says he doesn't have direct knowledge. But he does some thinking aloud for me. 'If you think about that point in time, Facebook had a million games, and they were all free. People would make these crappy games, and you have to authorise them to access your profile.' Accessing your profile also gave the company who made the game access to everyone you were connected to, and their profile information, just like Obama's election campaign.

'So, these were just data-harvesting apps, basically, and I think Facebook started to realise, "Oh, everybody's just pulling this data off Facebook." And that's the point where they recognised, "Hey, wait, why would we do this when we can keep the data to ourselves?" Facebook has shifted over time from, "You define who on Facebook you want to reach and we'll help you reach them," to, "You tell us some basic characteristics of the audience you want to reach, but we will actually target people to generate the most engagement."'

Therriault is sensitive to the accusation that overly targeted politics is manipulative and dishonest. But he also thinks that its powers over any individual are overstated. 'We are at the point where you're capable of tailoring messages to individuals. But it's very labour intensive. It's very messy. There's a bit of arrogance a lot of people have, to think that a campaign is going to try and manipulate them individually, as if taking over their mind is really the goal. The amount of work to brainwash everybody . . . Honestly, the way I explain it to people is, "Are you able to say, 'No, thank you,' to a college student knocking on your door, or get a piece of mail and throw it in the trash? Great, then you're immune. If you don't want to pay attention, you don't have to.'"

Now, I am inclined to agree that ignoring adverts is a core skill for survival in the modern world. When Twitter shows me adverts for beard oil, I don't go and buy beard oil, because I don't have a beard. I am easily swayed in unimportant things, like which biscuits to buy, but less easy to influence on important things like which way to vote. I think this makes me a normal adult. So, I don't worry about being psychologically manipulated. As we saw in the Introduction, when I did the Cambridge Psychometric Centre test, on which Cambridge Analytica claimed to base their work, social media activity isn't very accurate even on big categories like age and sex. I wouldn't bet money on any algorithm accurately predicting my mood and my vulnerabilities from what I post online.

But I do worry about the ease with which politicians today can give me one message, while showing a different message to somebody else, because I am receiving it in private, on a screen, instead of in the public arena. Andrew Therriault thinks that this, too, is overstated, and he thinks it pre-dates data science. 'I would go back a little further, to not just micro-targeting for individuals but moving from, say, three broadcast news stations to having cable news, and to a lesser extent, the increase in talk radio in the nineties. That all led to this idea that a campaign candidate can have different messages for different audiences.' Though even that idea is not new in political campaigning.

'Obviously, if Teddy Roosevelt, back in the early twentieth

century, went and talked to factory workers, versus going to talk to intellectuals, he's going to speak differently, that's a given. But doing that at scale has become much more of an art, to know your audience and tell them – not necessarily what you think they want to hear; it's more about what gets emphasised. A very clichéd one is politicians, when they're talking to male audiences, they're much more likely to talk about the economy, and foreign policy, but talking to female audiences are much more likely to talk about education and healthcare. That's a very classic example.' I saw exactly that during the 2019 General Election campaign in the UK. Facebook allowed people to see the various adverts that different political parties had run, and the broad audience they targeted. The Conservative party ran two near-identical ads promising that a vote for Boris would Get Brexit Done, but while men got that basic version, women seeing the ad got an extra detail: get Brexit done 'so we can invest in our NHS, schools and police'.

The other way that data science has transformed campaigning is that campaigns can track, in real time, which ads are working, and for which audiences. When we, the target voters, respond, or don't, we are giving them more data to refine their model. Of course, you can look at this as simply applying to politics the same tools that are already being used to sell us products. Like marketing, political campaigning has found new channels and media to get their message across, but the purpose has stayed much the same: to persuade us to do one thing and not another.

But in other ways, this shift from the megaphone to the micro-targeted message reflects a profound shift in the nature of politics over the past century.

*

In 1894, self-educated writer Benjamin Kidd published his first book, *Social Evolution*, while working as a civil service clerk. Looking back over the nineteenth century, he described the social and political impact of the industrial revolution: 'The worker is beginning to discover that what he has lost as an individual, he has gained as a class; and that by organisation he may obtain the power of meeting his masters on more equal terms.'[64] He believed that political equality for everyone was imminent.

The oldest of eleven children, Kidd grew up in Ireland and worked as a poorly paid clerk in London from the age of eighteen, until his book became a bestseller. Though sympathetic to the ideas of Karl Marx, Kidd thought socialism an impossible goal. What he wanted was a society where people would still compete, but in conditions of true equality. He saw thrilling potential in this new age of the masses.

Meanwhile in France, another writer was watching the 'era of crowds'[65] with gloomier eyes: 'Scarcely a century ago . . . the opinion of the masses scarcely counted and, most frequently indeed, did not count at all,' wrote Gustave Le Bon in *The Psychology of Crowds*. 'Today . . . the voice of the masses has become preponderant.'[66]

Le Bon didn't welcome this shift at all. He saw the goal of the masses as 'utterly destroying society as it now exists',[67] by limiting hours of labour, nationalising mines, railways and factories, and 'the elimination of the upper classes for the benefit of the popular classes'.[68] He was spooked by the 'spectre of communism' that was haunting Europe, according to the first sentence of Marx's 1850 *Communist Manifesto*. Europe and North America had seen revolutions overthrow kings and governments, from the English Civil War in the seventeenth century to the American and French Revolutions at the end of the eighteenth century. The year 1848 saw so many revolutions in Europe, that historians sometimes call it 'the Springtime of the Peoples'.

In April 1848, masses of people joined a huge Chartist demonstration on Kennington Common in London. From there, the leaders planned to march to Westminster Bridge, and across it to the Houses of Parliament. They had a petition to present with millions of signatures, calling for a vote for every man, secret ballots, and other reforms to make Parliament more democratic.[69] According to the *Illustrated London News*, it took a cart drawn by four farm horses to carry the signatures in 'five huge bales or bundles'.[70]

Frightened that Britain might be about to see a revolution like other European countries, the authorities fortified all the bridges across the Thames with soldiers, police and special constables to prevent the demonstrators from reaching Westminster or the City of London. The petition was finally delivered, but the government rejected it, arrested many of the leaders and suppressed the movement.

The fear that the common people, who outnumbered the upper classes, would organise themselves to overthrow their rulers was well founded. Gustave Le Bon had seen the short-lived Paris Commune of 1871 take over his home city, erect barricades, rule through an elected council and end in the 'bloody week' of executions by the victorious national army. But this suspicion of the masses was not confined to conservatives. In the late nineteenth century, many who saw themselves as progressive and liberal also saw the lower classes as a lower type of human being. It wasn't only non-Europeans, or people with different skin colours, who were regarded as a race apart.

Francis Galton, a cousin of Charles Darwin and originator of many important ideas in statistics, coined the word eugenics. Now synonymous with the Nazi's genocidal projects, in the late nineteenth and early twentieth centuries, the idea was very popular in educated circles. Galton gave a talk to the Sociological Society in 1904, calling for eugenics to be introduced into the national conscience 'like a new religion'.[71] Among the audience were many prominent cultural figures.

Writer George Bernard Shaw agreed that 'nothing but a eugenic religion can save our civilization from the fate that has overtaken all previous civilizations', calling for the institution of marriage to be separated from the selective breeding of humans. Author H. G. Wells was sceptical about Galton's call for the brightest and best to have more children, and instead suggested, 'It is in the sterilization of failures, and not in the selection of successes for breeding, that the possibility of an improvement of the human stock lies.'

Eugenicists like Galton, Shaw and Wells did not see mass politics as a positive force that would widen democracy and equality. They talked about ordinary people as passive products of genetics and environment, to be bred into improvement like farm animals – or out of existence altogether.

However, Benjamin Kidd rose to challenge the wave of enthusiasm in the room. Responding to Galton's speech, Kidd warned that eugenics 'might renew, in the name of science, tyrannies that it took long ages of social revolution to emerge from'.[72] Kidd accepted that evolution worked in humans as well as animals, but he believed the evolution of society allowed people to overcome the brutal selfishness of natural selection. With horrible prescience, Kidd suggested that 'many of our ardent reformers would often be willing to put us into lethal chambers, if our minds and bodies did not conform to certain standards'.

*

The masses didn't enter public life by invitation from those in charge. It was by clamouring outside the halls of power until those inside couldn't ignore the noise that they gained their part in political debate, and their share of power. 'The meaning of the twentieth century is the freedom of the individual soul,'[73] declared American writer W. E. B. Du Bois in 1915, in support of votes for women as well as Black men. To liberate each and every soul meant liberating all of them. Only through mass enfranchisement could the individual gain freedom and a political voice.

To demand the vote for all adults was to demand equal treatment as citizens, and an end to the exclusion of millions of people from political power on the grounds of their sex, or race, or how much property they owned. It was a demand to be treated the same as everyone else. That is why the struggle, not only for the formal right to vote, but for the ability to exercise that right, and for it to have some meaning, has continued throughout the twentieth century and to this day. In a democracy, the right to vote is the right to exercise your share in political power; to vote certain people into government, and to vote them out again if you don't like the way they govern.

The expansion of the vote changed the relationship between the masses and the government. Before the masses had a vote, governments needed only their tacit consent, or the absence of resistance. Now, governments have to convince a majority of the electorate to vote for them, or to vote against their rivals. They need to have a two-way relationship with the population they govern. Public opinion polls and surveys became an industry, as we saw in Chapter 2. That industry is driven not only by market researchers who want to sell products, but by government and political researchers who want to know what the electorate is thinking, and to predict how they might vote.

Where my mother grew up, in the northern English fishing port of Grimsby, almost everybody voted for the Labour Party. That was the party that claimed to represent the interests of the working class. Many of my mother's older relatives worked in the docks, which was brutally hard manual labour with little job security. Dock workers around Britain went on strike many times for better pay and conditions, including a national strike in 1970 that provoked the government to declare a state of emergency.

Today, the port of Grimsby employs hundreds, not thousands of people. The large working-class community centred on the docks no longer exists. In 2019, the town elected its first Conservative MP since 1945. In common with many other towns and cities in England, working-class voters in Grimsby have shifted their political allegiance, if they feel any loyalty at all. Across the nation, voters today are much less likely to say they identify strongly with a political party. Since British voters were first asked the question in 1987, the proportion of people who said they identified very, or fairly, strongly with a political party fell from 44 per cent to 35 per cent.[74] Among younger voters, around half said they felt no identification with a political party.[75]

Politicians can no longer see voters in terms of mass communities who can be assumed to lean in the same direction politically. That's one of the reasons they have turned to data to understand what makes us tick. But exercising power through the vote is not the only way that mass movements have changed the course of society.

*

American composer Aaron Copland produced ballet scores, symphonies and film music, but his best-known work is probably a simple piece for brass and drums that lasts barely three minutes. You have almost certainly heard it played at an Olympic Opening Ceremony, or in an arrangement by rock band Emerson, Lake & Palmer that made the charts in 1977. 'Fanfare for the Common Man' was commissioned for the Cincinnati Symphony in 1942.

'It was the common man, after all, who was doing all the dirty work in the war and the army,' Copland wrote later. 'He deserved a fanfare.'[76] Copland considered several titles before choosing that one, inspired by a rousing speech delivered by US Vice President Henry A. Wallace in 1942. Wallace declared: 'The century on which we are entering – the century which will come into being after this war – can be and must be the century of the common man.'[77]

Wallace had very specific reasons for invoking a common humanity, and a shared dedication to freedom. He was addressing an international audience, but also a domestic audience of Americans who had just joined the Second World War in alliance against Hitler and his allies, and would shortly be asked to support it both financially and by joining up to fight.

'Everywhere the common people are on the march,' he told

them, invoking a sense of history as progress; specifically, progress towards the Four Freedoms outlined by President Roosevelt in 1941: freedom of religion, freedom of expression, freedom from fear and freedom from want. Wallace praises what he calls 'a long-drawn-out people's revolution. In this Great Revolution of the people, there were the American Revolution of 1775, the French Revolution of 1792, the Latin-American revolutions of the Bolivarian era, the German Revolution of 1848, and the Russian Revolution of 1917.'

It's less surprising to hear an American politician praise all these revolutions when you remember that, at this point, Soviet Russia was an ally against Nazi Germany. Even so, the invocation of popular uprisings of the masses against their rulers is a bold move for a US vice president. Only the year before, Washington was braced for a protest march of a hundred thousand people, called by Asa Philip Randolph, President of the Brotherhood of Sleeping Car Porters.

Inspired by reading W. E. B. Du Bois's book, *The Souls of Black Folk*, Randolph founded a radical newspaper, *The Messenger*, in 1917, and began organising Black workers into trade unions. Pullman was the biggest single employer of Black men, who worked as porters – really, more like stewards – for his railway sleeping cars. That meant that, organised, the sleeping car porters could be a force for freedom and equality.

In an era when employment was highly segregated, workers' organisations sometimes saw other workers as a direct threat to their own pay and conditions. White Pullman conductors feared that they might be replaced by porters who were on lower rates. This, combined with outright racist attitudes, meant that many trade unions had few or no non-White members. The porters themselves complained that the company was hiring less skilled Japanese and Filipino workers, instead of paying experienced sleeping car attendants. Randolph saw clearly that this kind of racial division could only weaken labour movements. 'There can be no such thing as a colored labor union or a Filipino labor union,' he wrote in 1939. 'All unions are workers' unions, or should be ... the Brotherhood accepts all porters and attendants and maids and bus boys as members, regardless of race, color, creed or nationality.'[78]

Randolph pushed the American Federation of Labor to see that issues of racial discrimination, lynching, and the case of the Scottsboro Boys (which inspired the novel *To Kill a Mockingbird*)[79] were all causes that the wider labour movement should support. In 1941, Randolph called a national march to demand full racial integration in America's defence industry. The Brotherhood of Sleeping Car Porters carried word from city to city, mustering support through churches, universities and local community organisations. The National Association for the Advancement of Colored People (NAACP) mobilised their local chapters.

With America fighting for democratic values in a world war, a march for equal rights would be an embarrassing sign of social division at home. Six days before the march was due to bring 100,000 people to the Lincoln Memorial in Washington, President Franklin D. Roosevelt signed an executive order, *Prohibition of Discrimination in the Defense Industry*,[80] against any discrimination 'based on race, creed, color or national origin'. Since the immediate goal of the campaign had been achieved, Randolph called off the march – for now.

Randolph finally got his march on Washington over twenty years later. In 1963, Martin Luther King Jr delivered his famous speech,

'I Have a Dream . . .' to a crowd of over 200,000 people. The March on Washington for Jobs and Justice was organised by Randolph and the man he nicknamed Mr March-on-Washington, Bayard Rustin.[81] The gathering of people from different races, classes and organisations was the embodiment of mass movements, coming together to demand equal treatment for all: equal freedoms, equal rights, an equal voice in democracy.

'In a sense,' said Reverend King, 'we've come to our nation's capital to cash a check. When the architects of our Republic wrote the magnificent words of the Constitution and the Declaration of Independence, they were signing a promissory note to which every American was to fall heir. This note was a promise that all men, yes, black men as well as white men, would be guaranteed the unalienable rights of life, liberty and the pursuit of happiness.'[82]

But, said King, so far that promise had not been honoured, at least not for her citizens of colour. 'So, we've come to cash this check, a check that will give upon demand the riches of freedom, and the security of justice . . . Now is the time to make real the promises of democracy.'[83]

King pointed to the 'white brothers' who had joined the march, realising that 'their freedom is inextricably bound to our freedom'. One of them was Tom Hayden, later Senator Hayden: 'I came joyously from Harlem on a bus crowded with black people,'[84] wrote Hayden. 'It seemed certain on that magical day that our combined movements . . . would unite against racism, poverty, unemployment, and the deadly weight of the arms race.'

Tom Hayden grew up in a White suburb of Detroit, where his general sense of dissatisfaction with the world found expression in 'Beat' writers like Allen Ginsberg, and anti-heroes like James Dean in *Rebel Without a Cause*. It was Jack Kerouac's *On the Road* that inspired Hayden to hitchhike to California in 1960, an experience that kindled his lifelong political activism. On his road trip, Hayden interviewed Martin Luther King for the student newspaper he edited, met Latino farmworkers organising for their rights, and fell in love with his future wife, Sandra 'Casey' Cason, when he saw her give a rousing speech against segregation.

'I grew up in a time of alienated anti-heroes like Holden Caulfield and James Dean,' Hayden wrote in 2012. 'But after seeing the courage of black people in the south fighting Jim Crow laws, I was converted to activism.'[85] Hayden joined the Freedom Rides, was beaten up by the police and spent a short time in jail in 1961. While he was there, he began drafting a manifesto for Students for a Democratic Society (SDS), which was adopted in 1962 as the Port Huron Statement.

Hayden described the group that gathered at Port Huron, a conference centre belonging to the United Auto Workers trade union, as 'children who'd rejected the Old Left of their parents, black student civil rights activists seeking northern campus allies, children of the New Deal labor-left and student journalists.'[86] What is striking about the statement they agreed is its lack of demands.

When the Chartists gathered in 1848 to take their petition to Parliament, they were asking for distinct changes to the political system: a vote for all adult men, annual elections to Parliament, constituencies of equal numbers of voters, pay for MPs and secret ballots.[87] The Port Huron Statement[88] makes no such demands. It describes what is wrong with society in the students' eyes: not just poverty, unemployment and war, but its failure to meet their generation's emotional and psychological needs. 'We oppose the depersonalisation that reduces human beings to the status of things,' it declares. 'The goal of man and society should be human independence; a concern not with image or popularity but with finding a meaning in life that is personally authentic.'[89]

In some ways, its cry of outrage at the dehumanisation of humanity echoes manifestos going back to the Founding Fathers and beyond, calling for each individual to be recognised as a person with needs and desires beyond mere survival, and with the right to self-determination. But there's a subtle change of direction. Instead of the right to self-determination so that people can give their *own* lives meaning and purpose beyond individual survival, have an impact on the wider world and transcend the limitations of private life, the students declared that politics should be a 'means of finding meaning in personal life'.[90]

By participating in politics, the personal alienation of the individual could be healed. The new political order 'should provide outlets for the expression of personal grievance and aspiration . . . So that private problems – from bad recreation facilities to personal alienation – are formulated as general issues.'[91] The purpose of politics sounds almost therapeutic here.

The statement captures a moment at the beginning of a new political age. It talks about 'loneliness, estrangement, isolation',[92] and 'the felt powerlessness of ordinary people'.[93] It matters, not only that people lack power, but that they *feel* powerless.

Many of the themes in the Port Huron Statement come to fruition in the personalised politics we know today. The quest for personal authenticity has become a goal in itself. Demands to exercise a fair share of power in society are giving way to demands for recognition by those in power, not only of grievances, but of the emotions those grievances provoke. Disconnected from communities, individuals feel isolated and alienated. Meanwhile, political principles and allegiances are now felt so deeply within the self that to disagree with somebody can be like a personal insult or attack.

The distinction between politics and personal feelings had begun to fray as the sixties drew to a close, and a slogan emerged that captured this new kind of politics. 'The Personal Is Political' is the title of a short article written by Carol Hanisch, an activist in the Mississippi Civil Rights Movement, for a women's caucus. She later said that she didn't choose the title, but she articulates the Women's Liberation Movement's case that their focus on aspects of women's personal lives *was* political. 'Personal problems are political problems,' she wrote. 'There are no personal solutions at this time. There is only collective action for a collective solution.'[94]

Carol Hanisch was a founder of New York Radical Women, the group that protested at the Miss America beauty contest in 1968. They were all women who were active in other movements – Civil Rights, Anti-War or the New Left – who wanted to discuss politics without men present. After the meeting proper ended, they would continue arguing and questioning into the small hours at a nearby restaurant, over apple pie à la mode.

The Women's Liberation Movement took shape as women talked about issues that they had in common. Carol Hanisch described how women were leaving the New Left movements and apparently dropping out of politics. 'The obvious reasons are that we are tired of being sex slaves and doing shitwork for men whose hypocrisy is so blatant in their political stance of liberation for everybody (else),'[95] she wrote.

Hanisch wrote her article in 1969, responding to criticism

that Women's Liberation Movement meetings were just therapy for individuals dissatisfied with their personal lives, not politics. 'Therapy assumes that someone is sick and that there is a cure, e.g. a personal solution,' she wrote. 'I am greatly offended that I or any other woman is thought to need therapy in the first place. Women are messed over, not messed up! We need to change the objective conditions, not adjust to them.'[96]

Four demands to change objective conditions came out of the first British Women's Liberation Movement (BWLM) conference in 1970, held at Ruskin College in Oxford. They were:

1. Equal pay for equal work.
2. Equal education and equal opportunities.
3. Free contraception and abortion on demand.
4. Free 24-hour nurseries.

All of those demands required political changes that would affect individual lives. They are yet to be achieved in full in the UK, let alone worldwide.

Later conferences added further demands. In 1974, the Edinburgh BWLM conference added a demand for legal and financial independence for all women, because it was still common for banks to require a husband, or even a father, to sign loan agreements or be the main signatory on a bank account. The same conference called for the right to a self-defined sexuality, and an end to discrimination against lesbians.

The last national conference in 1978 adopted a seventh demand, which shifted the focus firmly onto men as the problem for women. 'Freedom for all women from intimidation by the threat or use of male violence, an end to the laws, assumptions and institutions that perpetuate male dominance and men's aggression towards women.'[97] Whereas any man could choose to support any of the first four demands – for childcare, contraception, equal pay and opportunities – it was much harder for a man to opt out of male dominance. Simply being a man cast him as the antagonist, whatever his beliefs about how society should be. No matter what he thought, said or did, what he *was* mattered more.

Divisions and disagreements over whether activism should focus on men and the patriarchy, or on using existing power structures to get women equal rights and practical freedoms, fractured the movement. From then on, the idea that Women's Liberation, or feminism, should be one organisation with shared goals, was abandoned.

*

A few years ago, I was touring a comedy show with a friend. Before you start imagining the glamour, this was three of us driving a van around the small towns of England, unloading our ludicrous props into a small arts centre, and performing to anywhere between thirty and five hundred local people. We'd been tipped off to arrive at this venue early because there was a very good Indian restaurant in the same building, which was an old military hut, among trees, slightly off the main road.

I had a strange feeling that this place was familiar and yet utterly unfamiliar, and as our poppadoms arrived, I realised why: I had been here before. Not the restaurant, or even the hut, but the site. The ex-military hut had once been part of Greenham Common Air Force Base, which was used as a storage facility for American nuclear weapons, and as a teenager, I had taken part in protests at the perimeter fence.

It felt quite odd to be on the inside of this building now, eating curry, when all those years before, there were men with guns guarding this place from women and girls like me, defending the nuclear weapons kept here, and making sure we did not get inside – though the women who lived in the peace camp outside the gates did get in, quite often, and none of them were ever shot.

I would like to claim that it was kids like me who tipped the balance and helped bring about the removal of those nuclear weapons by turning up once or twice at weekends to attach coloured wool to the fence. More likely, when the Cold War ended, there wasn't really any reason to keep the weapons or the base, but Greenham Common Women's Peace Camp embodied a change in politics during the late twentieth century.

The camp was started in 1981, when a small group of peace campaigners walked from south Wales to a small Royal Air Force base to protest against plans to site American nuclear weapons there. The marchers didn't have a plan for what to do when they got there, but a few of them had the urge to do what the suffragettes had done sixty years before, and chain themselves to a fence.

As a mark of how much had changed since the suffragettes campaigned for the vote, the UK's Conservative government had been led since 1979 by Margaret Thatcher, Britain's first female prime minister. Unlike some of the peace marchers, Thatcher saw no essential connection between being a woman and being against war. In 1982 she led Great Britain into war with Argentina over the Falkland Islands, a small territory in the South Atlantic that Argentina claimed as Las Malvinas.

Thatcher's brand of Conservative politics saw the self-reliant individual as the future. She felt the trade unions had too much power in society. In fact, she described trade unions as 'the enemy within',[98] with the Argentinian armed forces as the external enemy. In 1984, Thatcher backed the National Coal Board against the National Union of Mineworkers, generally regarded as the strongest of the trade unions, in a dispute over closing down coal mines. The mineworkers went on strike, but the employers and government were prepared, with large stocks of coal to keep industries and electricity generation running. The strike went on for over a year and became increasingly bitter. Even today, communities and families are divided over who worked and who was on strike in early 1985.

Challenging the power of the trade unions was part of a broader political emphasis on the self-sufficient individual. 'Who is society?' asked Thatcher in 1987, in an interview for *Woman's Own* magazine. 'There is no such thing! There are individual men and women and there are families.'

The Greenham women, muddy and smelling of woodsmoke, in their wellies and jumpers and close-cropped haircuts, were a mirror-image of neatly dressed and coiffed Margaret Thatcher. Their non-violent direct action strategy was based on each individual taking responsibility for herself and her actions. Instead of a top-down structure where a few leaders decided what to do, and others followed instructions, small 'affinity groups' got together, agreed by consensus what they would do and did it. Sometimes that involved lying down in the road to stop a cruise missile convoy leaving the base; other times, it was cutting though the fence to dance on top of the missile silos by moonlight. The lack of a central organisation made it hard for the police and military authorities to predict or counter the actions. The ability of a handful of women to defy military security was often an embarrassment to the government.

Greenham Peace Camp wasn't initially women-only, but after several discussions, a vote was taken, and the few men were asked to leave the camp. This would shape not only what happened there, but the ripples that the camp sent out for years afterwards. The initial decision seems to have been pragmatic: 'They didn't help with the food, look after the kids,' said Jill 'Ray' Raymond, quoted in the book *Out of the Darkness*.[99] 'They get drunk, they are violent and they won't do the washing up.'

Some of the women felt that non-violent direct action would be easier without men around, as situations were less likely to escalate. It's not a radical political position to say that situations tend to unfold differently with and without men – some venues employ female security staff for this reason. Using their superior strength to inflict violence against women breaches social taboos for most men, and they certainly don't gain any status by winning a physical fight against a woman. There was also another strand that emerged over the years of the women-only camp; a thread of thinking that connected the nuclear weapons inside the base to individual male violence that some of the campers had experienced in their personal lives, and to a general feeling that the system was rigged against them as women.

It's worth noting that 1981 was only six years after the Sex Discrimination Act was passed. British people were still more likely than not to agree that 'a man's job is to earn money; a woman's job is to look after the home and family'.[100] What the women were doing by leaving their homes to live in a permanent protest camp was very controversial at the time.

The target of the protest expanded from cruise missiles to the patriarchy, echoing the shift in emphasis that had taken place in the British Women's Liberation Movement. We can see in this a change of focus from a shared external aim – getting nuclear weapons off the site – to goals based on what we would today call identity. Certainly, it was an early example of *who you are* being as important as *what you want to change*.

The mantra that 'The Personal Is Political' came up again and again, but while women at Greenham were discussing what they should eat, how to share out the domestic work and what it meant to sleep with women instead of men, the meaning of the phrase evolved. Instead of somebody's personal life revealing political changes that needed to be made in the public realm, personal life could itself embody change: eating a vegetarian diet, lesbian sex or cutting off your long hair could be political acts in themselves.

As the camp grew, its existence as a *women's* peace camp became more important. In *Out of the Darkness* Josetta Malcolm, who has since identified as non-binary, described their experience of a women-only camp as being 'like therapy, it was like a political awakening . . . to just see how things linked up about patriarchy, and nuclear arms, and war. Actually, to be in a women-only space and then have your voice heard as a woman is so empowering.'[101]

It's not new for individuals to gain a sense of their own capabilities, their independence and agency, through taking part in political action. After a major dock workers' strike in 1889, in which 10,000 strikers won their demand for a basic wage of sixpence an hour, John Burns wrote of the effects of victory, beyond the new 'dockers' tanner' pay rate. 'Still more important perhaps, is the fact that labour of the humbler kind has shown its capacity to organise itself;

its solidarity; its ability. The labourer has learned that combination can lead him to anything and everything. He has tasted success as the immediate fruit of combination, and he knows that the harvest he has just reaped is not the utmost he can look to gain.'[102] Achieving their shared goal of a pay increase also expanded the dockers' sense of what else they could attain if they worked together. Longer-term goals of social change began to seem possible.

Shared failure, on the other hand, can be demoralising. Despite widespread support from working-class organisations and others on the political left – including the women at Greenham Common – the National Union of Mineworkers accepted defeat in 1985. Membership of trade unions in the UK fell from a high of 13.3 million in 1979 to 10.4 million in 1987.[103] This fall partly reflected rising unemployment and a shift away from the manufacturing industry towards service industries, where part-time and casual workers were less likely to join a union. But Thatcher's government also introduced a number of laws limiting what trade unions could do, especially in support of another union's actions, which limited the effectiveness of strike action in disrupting businesses. Some of these laws protected the rights of individual workers to ignore the instructions of their own union to join industrial action.[104]

Today, membership of trade unions continues to fall. Just under 6.5 million people belonged to a British trade union in 2021, most of them employed in the public sector.[105] A series of high-profile strikes for better pay since the pandemic suggests that those who are in unions are more willing to take industrial action: public support varies between, for example, nurses (two thirds of the public supported strikes) and university staff (just under a third of the public supported their strike). Four out of five British adults think trade unions are important to protect workers' interests,[106] but the majority of UK people in employment don't belong to a union.

Kevin Price joined his first trade union in 1974, and probably collected food for the families of striking coal miners in 1984. He has been an active member of the Labour Party in his native Cambridge for many years, a city councillor for ten years, and was elected by Labour Party members in the region as their candidate for Mayor of Cambridgeshire and Peterborough in 2017. He campaigns strongly on poverty and housing, but those were not the political issues that led him to resign from Cambridge City Council. Stating that he had never voted against the Labour Party group on the council, but that his conscience would not allow him to vote for a specific motion brought by the Liberal Democrat group, he stood down in October 2020.

In his resignation speech,[107] given over Zoom because of Covid restrictions, Councillor Price endorsed the motion's headline, 'Trans Rights are Human Rights,' praising the rights that the 2010 Equality Act gives to people with protected characteristics that include gender reassignment. 'No one should have a problem with committing to upholding and defending those,' he said firmly. 'I certainly don't.' The three sentences he could not endorse, however, were: 'Trans Women are Women. Trans Men are Men. Non-binary individuals are non-binary.' He expressed the view that those three sentences would send 'chills down the spine' of many women, pointing out that there are different views about whether trans rights clash with women's rights, and that the body of the motion required the council to take a public stance on the issue.

Fair enough, you may think. An elected politician finds he can't support his party on a contentious issue and, rather than vote against them, resigns his seat. But a campaign followed to also get Kevin Price suspended or fired from his job as a porter at Clare College, at the University of Cambridge. In a 2017 interview for student magazine *Varsity*, he confided, 'When I first got the job I thought – this is not gonna be for me – posh people coming in and looking down their noses at me. But it's just not like that, it really isn't.'[108] Porters do all kinds of essential jobs that keep the university running, from security to sorting mail, and they're often a first port of call for students needing help.

Accusing Kevin Price of showing 'a brazen contempt for the rights and dignity of trans and non-binary people', the Union of Clare Students said that his views would make trans and non-binary students feel 'unsafe in the community they are supposed to be protected by'.[109] In a statement condemning transphobia in the Cambridge Labour Party, Cambridge University Liberal Association pointed to tweets that Kevin Price had shared and liked.[110]

I'm not describing this case to ask you to take sides with or against Kevin Price. I know people hold strong and wildly varying views on the issue of trans rights, what exactly they entail, and how, if at all, they are in tension with the rights of women. What I think this story illustrates is how much the nature of politics has changed in living memory. It demonstrates how much of what is at stake here is about how people are described and seen by others. The motion over which councillor Price resigned commits the council to 'stating publicly that trans rights are human rights', alongside other symbolic commitments.[111] A few of these could have practical consequences, but on the whole, those are already covered by laws such as the Equality Act, cited by Councillor Price.

The old style of politics, in which trade unions exist to protect the employment rights of workers, is now in direct conflict with the new style of politics. The personal really is political, in the sense that individual behaviour and language are now political issues. The right of a trans woman to be called a woman, of a trans man to be called a man, or a non-binary person to be called non-binary, is now so important that student organisations feel justified in calling for somebody to lose their job.

Identity has become a common theme in politics today. Sometimes that's obvious, in issues like the campaign for self-identification for trans people, literally the right to be recognised by society, the law and others the same way you see yourself. At other times, it's a new way of seeing existing issues. Britain's vote to leave the European Union, the US election of Donald Trump as president and the political impasse in Northern Ireland have all variously been explained by the drive to find recognition and a sense of belonging; to have your identity recognised and respected.

It is true that political decisions and allegiances are based on more than economic self-interest. Surveys have shown that, in most Western democracies, for every person who agrees that 'most elected officials care what people like me think', two people disagree. That means that two thirds of the voting public think that most elected officials *don't* care what people like them think.[112] They feel voiceless, invisible, ignored, like the students who wrote the Port Huron Statement.

The question that explanations grounded in identity often leave out, though, is: 'Who has the power?' The struggle in the 1980s between the National Union of Mineworkers and Thatcher's Conservative government was not about coal. It was about who had power in the country. The collective economic and social power of the trade unions was a challenge to the power of the elected government. The Greenham Common women, on the other hand, explicitly rejected systems that centralised power. They wanted to erode, if not the physical power of guns and missiles, then the legitimacy of holding that power. They wanted to be visible, and to render visible aspects of themselves and their lives that they felt had been ignored or devalued. They wanted their grievances, their experiences, and their existence, to be recognised.

For most people, their employer has economic power over them, because they depend on earning money to live. Laws protecting people from discrimination in getting a job, or from being arbitrarily fired without good reason, counterbalance that power. Belonging to a trade union may also offer some collective power, if enough employees are prepared to stick up for one another. But, unless you have a really specific set of skills or knowledge that your employer can't do without, there's always a power imbalance.

That's why I am very wary of workplaces that promote equality by telling employees in detail how to behave. Insisting that all staff members use certain words and not others, or wear badges announcing that they support a political cause, lets a company look very moral and principled. It's certainly a lot cheaper than providing free twenty-four-hour childcare, for example.

Most of us can think of a past boss, or colleague, whose behaviour crossed the line between mere annoyance and outright discrimination or harassment. Many of us have wished that our employer would do something about the person who is making our working life difficult. But once you have set the precedent that an employer can discipline your colleagues, or even sack them, for saying the wrong words, it's very hard to draw new lines and say, 'But not THOSE words!'

After all, if people's feelings about themselves are reason enough to call for somebody else to lose their job, who is to say which feelings should be protected that way? What if you have religious feelings which are hurt if I come to work in a sleeveless dress that shows my knees? What if you have strong vegan beliefs and would be upset if I brought a sausage roll in for my lunch?

How much power do you want to give your employer to decide what words and behaviour are acceptable? Should somebody lose their job for holding opinions that you find uncomfortable or offensive? Would you be happy to have the same rules applied to you?

*

On 26 May 1953, Aaron Copland, composer of the 'Fanfare for the Common Man', was called before Joseph McCarthy's Senate Committee, to be questioned about his political loyalties and past activities. One of those past activities was contributing, along with fellow composer Leonard Bernstein and others, to a benefit concert for Hanns Eisler, a German-Austrian musician who had fled Germany in 1933 when the Nazis banned his music. Eisler was deported back to East Germany in 1948 after his own sister denounced him as a communist.

Copland had been associated with communist organisations and events in the past. He had even written a song, 'Into the Streets May First',[113] published in the *Workers' Song Book* in 1935. But since that time, he had been employed as a lecturer by the US Government and sent abroad to speak about American music.

The committee asked Copland a lot of detailed questions about letters and petitions he had signed, meetings he had attended, and people and organisations that he appeared to support. Copland said this support was mostly confined to 'signing my name to a protest or statement, which I thought I had a right to do as an American citizen'.

'You have a right to defend communism or the Communist Party, Hanns Eisler . . . or anything else,' replied the Chairman. 'You have a perfect right to do it, but the question is why were you selected as a lecturer when you exercised that right so often?' The committee went on to ask Copland whether a communist should be allowed to teach in American schools. Copland had thought about calling his 'Fanfare for the Common Man' either 'Fanfare for Four Freedoms' or 'Fanfare for Democracy'. He wasn't feeling much freedom of expression, or freedom from fear, on that Tuesday in May, in 1953.

I could point to the similarities between Copland's questioning and the calls for Kevin Price to lose his job. There are echoes in the raking back over old statements of support, in the implication of guilt by association, and in the suggestion that people holding certain views should not be in certain professional roles, but I think what's more interesting are the differences.

Senator McCarthy was very clear about the problem with Copland and the other people on his list. The problem was that they were, or he believed they were, communists, and communists wanted to get rid of the American political system and replace it with communism. They were on the side of the Soviet Union, against which America was fighting a cold war. McCarthy was on the side of the American government and, in his eyes, freedom and democracy.

They could exercise their democratic rights, said McCarthy, but if they exercised them against American interests, they were not entitled to employment, especially in jobs where they might push their communist ideas onto others. McCarthy was using the power of the government of one of the world's superpowers in a struggle between two rival political systems that saw the world in very different ways. He didn't care whether the person in front of him was Aaron Copland or Kirk Douglas. It wasn't personal. It was about who had power.

The student organisations who wanted Kevin Price out of his job at Clare College also objected to his ideas. They claimed that somebody holding those ideas should not be in a position where he helped students, because that would make some of those students feel unsafe. Not because there was any suggestion of physical threat, but because he might not *see* them the same way they see themselves. There was a danger that trans or non-binary students would not see themselves reflected in Kevin Price's eyes as the person they felt themselves to be. And that jarring dissonance of perception was felt as a form of violence to their personhood. It was personal. It was about people's feelings, and what protection those feelings deserve from those with power.

A few years ago, a friend of mine got a letter from somebody she worked with. The letter declared deep feelings of love for her, which her colleague had never declared before. In a romcom, the next scene would be him in his flat, looking at all the scribbled-out drafts of the letter, his agonising interrupted by urgent ringing on the doorbell. He would open the door to find her outside, holding the letter, probably crying. If it was a British romcom she'd be soaked by the pouring rain. She'd fling herself into his arms and say she'd never known that he felt that way, and why hadn't he said anything sooner, and the next scene would be them on honeymoon.

But this wasn't a romcom. She did know he felt that way. It had been obvious for ages, but she didn't feel the same way about him, and they were working together, so she had carefully avoided giving him any reason to think his feelings were welcomed or returned, or even noticed. Until now, when (as the letter explained) his therapist had told him that it was vital for his emotional healing that he should declare his feelings. I expect he felt much better as soon as he had posted the letter.

But what about my friend? What was she supposed to do with this? She couldn't go on pretending not to notice that her colleague was in love with her. She was not in love with him. Did the therapist give any thought to the impact on other people of this therapeutic declaration? Did they discuss the difficulties of working together that might, and in fact did, follow?

Being human means living and working with people whose feelings are different from yours. The fact that you feel a particular way about them, or that you want them to feel a particular way about you, puts them under no special obligation to you, beyond the normal civilised behaviour they owe to everybody.

Human beings want to belong, and to be respected and have our human dignity recognised by others. This is not new. What is new is the importance that society in general gives to these individual wants. Mass movements of the late nineteenth and early twentieth centuries were largely demanding to be treated the same as other people. Men without property, women and Black people wanted equal voting rights. Black citizens wanted equal access to jobs, education and public spaces. Women wanted the same rights as men to financial independence, and to paid work outside the home, regardless of whether they were married or had children. Even demanding particular rights, like the right to choose the termination of an unwanted pregnancy, was seen as a step towards equality. If women can't control whether, or when, they have children, they have less control over their own lives than men do.

Today, many movements demand not to be treated the same as others, but to be treated differently. This isn't necessarily a bad thing. Because people *are* different, equal access to public goods like transport, education or democracy can mean meeting different needs. An eighty-six-year-old probably needs taxis, not cycle lanes. A blind person can't read printed election leaflets stuffed through a letterbox. If you don't have that particular experience, it may not have occurred to you that those things could be vital to somebody else. But if you, as a matter of principle, think that equality and fairness are important, you understand that other people should get things that you don't need yourself. Not until you're eighty-six, anyway.

It's also obvious that we don't all want the same things. I remember, as a student, some socialist group telling me that after the revolution, everyone would have 'steak and potatoes every night'. I was a vegetarian at the time, so that wasn't a very appealing slogan. The point is that everyone should be equally free to have steak and potatoes, or kosher chicken, or lentil curry, or whatever flavourless protein shake the tech bros are currently swigging. And that equal freedom is something that we can *all* campaign for, because it's in everybody's interest to be free to choose for yourself.

Having different experiences, opinions, needs and desires is not a problem. The problems arise when people's lives and experiences are seen as incommensurate. The idea that I can never imagine what it's like to be you, and so I'm not entitled to have an opinion on your political ideas, makes democracy impossible. If we can't find shared ground on common principles – equality, freedom, democracy – all we can do is make claims based on how strongly we feel about something, whether that something is respect for our religious beliefs or the fate of polar bears.

What matters to all of us tends to be more personal today – not only issues that are entangled with our everyday lives, our feelings or how we see ourselves: even geopolitical issues or which political party we vote for are taken much more personally. Earlier, I said that the number of people in the UK who identify strongly with a political party has fallen, and that's true. But there's a curious counterpoint to that. People are less open to social connection with those who disagree with them politically. This is true in the UK and the US.

A survey in 2017 found one in seven Republicans, and one in five Democrats, had no close friends who voted for the other party.[114] Similar surveys in the UK in 2017 found around four in five voters would not be happy if their child married somebody who voted for the other main party.[115] Feelings around Brexit in the year following the referendum ran almost as high among those who voted to Remain in the EU: around three quarters of them would not be happy if their child married a Leave voter. Leave voters were more tolerant, but still over half said they would not be happy to have a Remain voter marry into the family.[116]

Both party and political loyalties also affected how voters see others. We are more likely to see those who disagree with us politically as closed-minded, selfish, hypocritical, immoral or lazy, and less likely to call them intelligent or honest.[117] Negative feeling towards the other side, politically, has been steadily increasing since the 1980s.

The political has become personal in the sense that we take disagreement very personally. Yes, social media makes it easy to fire off a hasty reply in a few words, which tends to turn up the heat, but let's not kid ourselves. The quick satisfaction of venting an emotional response, and hopefully provoking emotions in the other person, is easier than reflecting on how we might win them over to our point of view. Politics is turning into a form of self-expression, rather than an arena to which we all bring our conflicting visions of how the world should be, and try to persuade others to join us in working towards ours, or even one in which we might listen to other people's ideas and change our minds.

Let's turn back to where we started this chapter: the data-driven campaigning that has abandoned the megaphone for the personalised message. Sasha Issenberg called Obama's campaign a break with the old ways of tracking public opinion, which relied on samples to represent the whole population. Now, freed from demographic and geographical categories, 'the electorate could be seen as a collection of individual citizens, who could each be measured and assessed on their own terms'.[118] Although the presidential campaign reduced every American to numbers, 'those numbers somehow captured the individuality of every voter'.[119]

You could see this as technology driving a move away from relating to voters in the context of their neighbourhood, their common experience with others like them, or the causes and ideals they share. But in an age when voters want to be treated as individuals, and in which demographics no longer reliably predict how you will vote, I see technology as meeting the campaigners' need to connect with the electorate, when they don't know how else to do it.

At the end of my conversation with Andrew Therriault, I asked him if he thought politics in general had become more personal in the last fifty to one hundred years. 'Yeah,' he said. 'From both the voters and the candidate side it's become more individualised, in terms of even the basic facts we're working with.' He blamed the changes in news media since everyone shared a few television channels and newspapers.

'Now, you think about talk radio, and the internet; suddenly there's much more insular communities. Politics is getting personalised in a way where it's not just a matter of people being given a set of objective facts and each person gets to make the decision for themselves. There's sort of . . . different realities. This is my worry watching US politics right now. I'm not worried about the algorithms that are saying what pieces of campaign mail you should get. I'm much more worried about the algorithms that say, "Here are the news stories you should be reading, and here are the YouTube videos you should be watching." That's where I get worried.'

Therriault ponders whether he is having nostalgia for a golden age that never existed, in which everyone shared a common world view, free of misinformation. I think there never was such a golden age, and also that the age we're in is not as polarised as we sometimes think. But I do think he is onto something when he looks at the personalisation of politics using data as a small part of a much bigger picture . . . or perhaps of many overlapping pictures.

'It's the personalisation of reality,' he says. 'This idea that your political reality can be whatever you want it to be, and whatever someone else wants to tell you it should be. And that there does not seem to be that sort of common grounding that we once had when things were less personalised.'

Now we have a sense of the hyper-personalised world we're living in, and how it emerged from the mass society that preceded it, I want to go back further in time. The next four chapters will trace three strands of history that brought us here.

If you asked me *how* the world is personalised, I would answer: with technology. So that's where we'll start, with the surprisingly deep roots of the cunning devices and software that we take for granted. But this is also a story of ever-widening choice, which makes personalisation both possible and inviting, so the following chapter will tell the complex and sometimes contradictory story of our increasing freedom to choose. Finally, we will come to our changing sense of who we are, and how our obsession with having our own identities reflected back to us came out of centuries of European history, philosophy and psychology.

Chapter 4: From File Cards to Profiles: Technology

'Philosophy is written in this grand book, the universe, which stands continually open to our gaze. But the book cannot be understood unless one first learns to comprehend the language and read the letters in which it is composed. It is written in the language of mathematics.'[120]

GALILEO GALILEI, 1623

In Galileo's time, there was no such thing as science. The observation of the natural world in an organised way, to understand the deeper causes of why things happened, was called natural philosophy.

Galileo, born in 1564, the same year as Shakespeare, was threatened with torture by the Roman Catholic Inquisition for believing the evidence of his telescope (that the Earth moves around the Sun, not vice versa) above the teachings of the church. He stood at a turning point in European thought between the authority of faith and tradition, and the questioning of all authority through experiment, observation, measurement and, above all, reason. Using mathematics to organise the world – both natural and human – conceptually and practically, would transform our lives over the next four centuries.

People have been counting, measuring and calculating since

before writing was invented. What was new to Europe in that era of philosophical, political and economic revolution was the idea that, using mathematics, human reason could extract the hidden secrets of nature. Far from setting limits to what humans should know, God had made creation intelligible to us, if we only understood its language, using not just the pure abstraction of mathematics, but the messier craft of statistics.

Statistics is a word with two meanings today. It can mean the numbers, or data, collected from the world, and it can also mean the methods used to make sense of those numbers, to tell a story about the world. Today's data collection and algorithms are essentially automated versions of the same thing, done by machines, faster and on a greater scale than any human could do them. But most of the ideas behind how they work are years, even centuries, older than the machines themselves.

Galileo lived long enough to meet John Milton in 1638. Milton would go on to write *Paradise Lost*, argue for the abolition of the Church of England and the monarchy and, as a member of Oliver Cromwell's Commonwealth government, defend the execution of the English King Charles I. In those turbulent times, the revolutions were political, religious and social, but also scientific and statistical.

Samuel Pepys records in his diary a visit to the Crown Tavern in 1668 with 'Captain Graunt telling pretty stories of people that have killed themselves'.[121] John Graunt was the man to tell these stories. A self-taught scholar, in 1662 he published *Natural and Political Observations on the Bills of Mortality*. The Lord Mayor of London collected regular records of plague deaths, and by 1625 the weekly *Bills of Mortality* had their own printing press.[122] The bills gave the numbers of deaths, by cause and parish, and also of christenings.

Pepys himself bought a copy of Graunt's book, which was impressive enough to get its author, a humble haberdasher, admitted to the newly formed Royal Society. That independent organisation of 'natural philosophers' still has the same motto – *Nullius In Verba* ('take nobody's word for it')[123] – though it now calls its members scientists.

Graunt's search for patterns and causes in the thousands of human lives and deaths recorded in the bills laid the foundations for statistical methods still used today. For example, he noted that some causes of death seemed to exact a steady toll, year after year. Deaths from 'chronical diseases . . . Consumptions, Dropsies, Jaundice, Gowt, Stone, Palsie, Scurvy, rising of the Lights, or Mother, Rickets, Aged, Agues, Feavers, Bloody-Flux, and Scowring',[124] as well as accidents and suicides, varied only a little around what we would call an average number. Smallpox, measles and plague, by contrast, came and went in waves, killing tens of thousands in a bad year.

He invented the Life Table, a chart of how long people lived, or what proportion could expect to survive to a given age. Only one in a hundred lived to the age of seventy-six in Graunt's London. Today, more than one in seven babies born in London can expect to live to a hundred.[125] Life Tables have become more accurate since Graunt's time, and they're still vital for anyone trying to predict the numbers of people who will need pensions, hospitals, special film screenings with free biscuits, and everything else that comes with old age.

It seems remarkable to modern eyes how much of Graunt's book was truly novel in its day. Large volumes of data had only recently become available, and in Graunt, these numbers found a curious mind. He believed it was his duty to decipher the book of the universe and find the underlying order, which would show God's purposes. He was the first to quantify how many of London's children never saw their sixth birthday – over a third – including his little daughter Susan who died in 1643.

Graunt himself died at fifty-three, penniless, after losing his home and business in the Great Fire of London, but his work inspired and informed the collection and use of statistics across Europe. Johann Peter Süssmilch, in his 1741 book *Divine Order*, compared Graunt to Christopher Columbus, daring to go where nobody else had gone before.[126] A pastor in a poor district of Berlin, with large numbers of textile workers employed in the new factories, Süssmilch looked for statistical evidence of God's order in the chaos of the newly industrialising cities of Germany.

Süssmilch had caught the zeitgeist and *Divine Order* was a bestseller. He was cited in a flurry of theological writings, all using statistics and science to show how God's plan was working out in the natural world. My favourite is Johann Gottfried Richter's *Icthyotheologie* or '*fish theology*' of 1754, which 'sought to lead men to admiration, reverence and love of God through the observation of fish.'[127]

Süssmilch showed that people in cities died faster. Working not only from numbers, but from his first-hand experience as a priest and a member of the 'Royal Directorate of the Poor', he saw the benefits of industrialisation to prosperity, but also the costs in human lives. We can see, in these early examples of statistics, the tension between a new ability to study human life as part of the natural world, subject to its laws of cause and effect, and a new recognition that each person in the mass is an individual human being.

'Given that the free will of humans has such a great influence on marriages, on the births that result from these, and on dying, it would seem that there is no rule to which these events are subject and according to which one could calculate their number in advance,' wrote philosopher Immanuel Kant in 1784. 'And yet the relevant statistics compiled annually in large countries demonstrate that these events occur just as much in accordance with constant natural laws as do inconstancies in the weather.'[128] Kant, another Prussian native, grappled all his life with this tension between human free will and natural laws. How could statistics reveal regular patterns in human life, when each individual decision is freely made by an individual human?

Belgian statistician Adolphe Quetelet first coined the term 'average man' (l'homme moyen) to represent the theoretical human, stripped of all individuality and random chance, in whom social laws could be seen at work. 'It is of primary importance to keep out of view man as he exists in an insulated, separate, or in an individual state, and to regard him only as a fraction of the species,' he wrote in 1835. 'In thus setting aside his individual nature, we get quit of all which is accidental, and the individual peculiarities, which exercise scarcely any influence over the mass, become effaced of their own accord, allowing the observer to seize the general results.'[129] By observing people as a mass, Quetelet hoped to deduce the social laws which drove what initially appeared to be individual decisions. 'It is society that prepares the crime; the guilty person is only the instrument who executes it,' Quetelet asserted. 'His crime is the fruit of the circumstances in which he finds himself.'[130]

Nevertheless, Quetelet identifies as one of man's 'noblest attributes' the ability to understand and modify the causes of social phenomena. 'As a member of the social body, he is subjected every instant to the necessity of these causes, and pays them a regular tribute; but as a man, employing all the energy of his intellectual faculties, he in some measure masters these causes, and modifies their effects, thus constantly endeavouring to improve his condition.'[131]

Quetelet also makes a clear distinction between a general law that applies to 'average man', and trying to predict the fate or actions of a real, specific person. 'Every application which one might attempt to make to a man in particular, must be essentially false, in the same way as if we were to pretend to determine the precise period of a person's death by looking into the tables of mortality.'[132] This distinction between making population-scale statistical predictions, and imagining that an individual's destiny is determined, continues to be poorly understood and applied to this day.

Quetelet's book, *A Treatise on Man and the Development of His Faculties*, consists largely of tables of numbers from which he draws conclusions about the influence of age, sex, weather and geography on anything from mortality to suicide. He finds the same shaped distribution of numbers around an average in all sorts of human qualities, including height and 'propensity to crime'.

The man who named Quetelet's ubiquitous curve the 'normal distribution' was Francis Galton, the cousin of Charles Darwin we met in the previous chapter, arguing that eugenics should be the new religion. Today, Galton's legacy is clouded by that association, and it's hard to separate his curiosity about mathematically analysing every aspect of human life from his conviction that the human race must harness eugenics to improve itself.

Galton studied medicine at Cambridge, before turning his

attention to the question that would obsess him: why do some families (namely, *his* family) produce so many eminent men? His book, *Hereditary Genius*, first published in 1869, outlined his theories, and how he had arrived at them by applying Quetelet's ideas about height to mental talents, and Darwin's theories of natural selection to different sections of humanity. His assumption that natural aptitudes and characters varied between 'races' applied both to groups he cheerfully labelled as Negroes, Irish Celts, Huguenots, Hebrews and Chinese, and to the different classes within one country.

Setting up his Anthropometric Laboratory at the International Health Exhibition in 1884, Galton charged people a small fee to have various things measured – sight and hearing, breathing capacity, reaction time and strength, as well as height and weight. Galton kept a copy, along with other dimensions including hair and eye colour, birthplace, age, sex and occupation, but no names.[133] One question led to a useful breakthrough: if qualities like height (or eminence) are truly inherited, why don't tall parents have ever taller children, for example?

The answer, Galton concluded, was what we now call 'regression to the mean', the tendency for extreme examples to return to a more average value. If we think of each generation as a distribution of heights, that distribution would always tend back towards what Galton called the 'racial centre' – what Quetelet might have called 'average man' for that particular population. In other words, your own parents' specific heights would predict your own height to some extent, but not completely. Regression, a measurement of the relationship between two sets of data, is still the backbone of using data to make predictions.

We will now leave Galton with the other statisticians in London, and cross the Atlantic to find the origins of modern computing. The Computer History Museum, started in Massachusetts in 1974, moved to Silicon Valley in 1998.[134] Its home for the last twenty years is an airy building surrounded by trees in Mountain View, close to Stanford University campus. When I visit on a sunny summer afternoon, I'm excited to find Hollerith's Desk on display.

Herman Hollerith's 1889 patent for 'certain new and useful Improvements in the Art and System of Computing Statistics' is a milestone in the history of technology, and this 'desk' is a replica of the equipment detailed in his patent.[135] In 1880, the US Census was struggling to turn all the data it collected on America's growing population (fifty million and rising) into information it could use. After eight years of processing data, the Census Bureau realised that, at this rate, the 1890 census report wouldn't be published till after 1900, and announced a competition to find a better way to do the job.

Former census employee Herman Hollerith won the competition with his ingenious system using punch cards. His method was the fastest at both processing the data from the completed census forms and tabulating it by categories: age, race, occupation, and so on.[136] Born in 1860, Hollerith was an engineer who began experimenting with mechanical data processing and, encouraged by medical doctor and statistician John Shaw Billings, adapted the file card system used by libraries.

Punch cards weren't a completely new idea. The mechanical Jacquard loom was already using them to control factory-woven patterns. The binary hole/no hole unit of information translated easily into a mechanical instruction, because the thread either could or could not pass through at that point of the pattern. Recording the census data on punch cards wasn't so revolutionary. It was quicker than writing a number or letter code, but a human still had to physically punch a hole in the right place on that household's census card. The real revolution was automating the reading of the punched cards, and that is where Hollerith's Desk came in.

Above the flat part of the desk are four rows of ten round dials, each the size of an alarm clock face, with a long and a short hand. On the right side of the desk is a device with a wooden handle that pulls down a brass frame, like a small printing press or a large kitchen implement for moulding waffles. Pushing the lever down presses together two brass plates.

This device could read the punched cards. Every time the operator pressed the handle down, metal pins passed through holes in the lower plate into cups of mercury, completing electrical circuits, but only if the punch card also had a hole in the same position. Otherwise, the electrical circuit stayed broken. Hollerith's device translated the hole/no hole binary into electricity/no electricity. This electrical signal, in turn, made one of the dials above turn, advancing the dial one position, adding one to the total.

Unlike a human clerk, the machine could read and count all the categories on the card simultaneously, and ring a bell to signal to the operator when it was ready for the next card. Using the machine, an operator could process and sort 7,000 cards a day.

Hollerith established the Tabulating Machine Company in 1896 to provide what we'd now call data-processing services to the US Census Bureau and other customers, including the Russian census and several European governments. By 1924, he had merged it with other companies to form the Computing-Tabulating-Recording Company, and retired to keep Guernsey cattle. New company president Thomas J. Watson Sr changed the name to International Business Machines – IBM.

Punch cards became the standard method of storing and transferring data, though punching and reading them became faster and more accurate as electronic machines took over the laborious punching and sorting that had previously been done by humans. Metal brushes and bars or rollers took over from pins dipping into mercury. Keypunch machines allowed operators to type with a conventional keyboard, and translated letters or numbers into correctly positioned holes.

In the mid-1930s, IBM's machines found a new market. America was struggling under the weight of the Great Depression, and in 1935, President Franklin D. Roosevelt signed into law the Social Security Act, providing his citizens with some hope for the future: 'The civilisation of the past hundred years, with its startling industrial changes, has tended more and more to make life insecure. Young people have come to wonder what would be their lot when they came to old age. The man with a job has wondered how long the job would last.'[137] By collecting money directly from workers' wages, the US Government would build up a fund from which to pay out benefits to the unemployed and elderly.

This new system threw up a new problem – how to keep track of who had paid in and who was entitled to draw out? By the start of 1937, when payroll taxes would first be collected, thirty million Americans needed to be registered with a unique nine-digit number. The US Post Office took on the task of distributing and collecting registration forms and allocating numbers, but how could the central Social Security Board (SSB) organise all this information?

Enter IBM. Thanks to the Great Depression, they had far more punch-card machines than they could sell to private companies. They provided over a thousand keypunch machines to create the punched record cards, and hundreds of accounting machines to process the cards. Every day, 600,000 cards were processed at the SSB's headquarters in Baltimore, Maryland.[138]

Canny Watson saw another opportunity. For employers, the new accounting responsibilities meant they were also in the market for IBM machines, to put their own payroll and accounting systems onto punch cards. His business model of leasing machines and selling file cards to feed them meant a steady flow of income. By 1939, the company had plants in England, France, Italy, the Philippines and Germany,[139] and subsidiary companies all over the world. IBM was close to having a worldwide monopoly on the major information-processing technology of its day.

In the IBM section of the Computer History Museum, a rousing chorus of 'Onward, ever onward!' rings out. It is IBM employees, literally singing from the company song sheet. In fact, there was a whole company song book, a generous system of rewards for successful salesmen, rules of conduct (no alcohol during business hours) and a company motto: 'Think!'

The song is so catchy that for days afterwards I find myself humming, 'Ever onward, IBM!'

One item in the IBM section of the museum brought me up short. All credit to IBM, who sponsor the museum's work, for acknowledging this part of the company's history, I thought. It's a photograph of Thomas J. Watson Sr, in Berlin on 28 June 1937, taking tea with Adolf Hitler.

The role that IBM's German subsidiary, Dehomag, played in

the Third Reich was researched by Edwin Black and described in appalling detail in his book, *IBM and the Holocaust*. It is the most extreme example I can imagine of how data-gathering and processing technology can be used for evil as well as for good.

Dehomag was older than IBM. In 1910, German adding machine salesman Willy Heidinger bought the licence to use Hollerith's technology and created Deutsche Hollerith-Maschinen Gesellschaft – Dehomag. Hollerith's company got a share of Dehomag and also received royalties. This business relationship survived the First World War, but not Germany's post-war hyper-inflation. Heidinger was unable to pay the royalties, and in 1922, Watson travelled to Germany to deliver an ultimatum that left IBM owning 90 per cent, with Heidinger still in charge on the ground.

In 1933, within months of Hitler coming to power in Germany, Dehomag had won a huge contract to run the census that would register all Germans, sorted by age, sex, occupation, religion and race. That census was only the beginning of a regime that would track everything in the expanding territory of the Third Reich via IBM's punch cards: buildings, raw materials, money, weapons, food, vehicles, horses, cows, railway rolling stock and, of course, human beings.

Many countries and US states passed eugenics laws in the early twentieth century. Not only scientists but governments across the world saw the selective breeding of people as a desirable goal. Hundreds of thousands of people were sterilised without their consent. The first American state to pass a compulsory sterilisation law was Indiana in 1907, and the last American state to repeal its sterilisation law was Oregon in 1983.[140] Denmark, Sweden, Norway, Finland and Iceland all had laws in effect until the 1970s.[141] So, when Hitler passed the Law for the Prevention of Genetically Sick Offspring, in January 1934, he was not completely out of step with European or American government thinking. The compulsory sterilisation programme began with mental and physical disability or illness, and rapidly moved on to people displaying the wrong attitudes or 'anti-social' behaviour.

By the time Watson took tea with Hitler in 1937, it was very clear that Hitler's Germany was a dictatorship pursuing a specific and brutal policy against Jewish people, and other 'undesirable' groups. On the same 1937 visit, Watson accepted Hitler's medal, the Merit Cross of the German Eagle with Star, for 'foreign nationals who have made themselves deserving of the German Reich'.

The Reich created a special unit within the War Ministry to control all punch-card technology in Greater Germany. The Maschinelles Berichtswesen, MB, would eventually assume direct oversight of Dehomag. So central was IBM's punch-card technology to the Reich that concentration camps had their own record offices, and sometimes their own Hollerith machines, tracking each person's fate. Codes on the cards distinguished different classes of inmates – homosexuals, political prisoners, Jews, and so on – any work skills, and their eventual fate. Only those who were murdered immediately on arrival did not enter the system.

In June 1940, after the Nazis had annexed Austria and invaded Czechoslovakia, Poland, Denmark, Norway, Holland and Belgium, Watson wrote to Hitler to return his medal. While this played well at home, it was the beginning of tense negotiations between IBM, Dehomag and the Nazi government. The Reich suddenly saw how utterly dependent it was on this all-pervading technology, and how much intimate knowledge IBM potentially had of almost everything the Reich was doing, or planned to do in future.

The ultimate outcome was satisfactory to Watson: American IBM retained majority ownership of Dehomag, but gave up detailed knowledge of its operations. This would later enable Watson to collect machinery and other assets in post-war Europe, while denying knowledge of the uses to which it had been put.

Meanwhile, IBM in America was indispensable to the US government – not only to the American Social Security system, but also for armaments, personnel, transport and money, which would all be tracked and recorded via IBM's ubiquitous punch cards. The American census of 1940, using IBM technology, also enabled the US Government to identify Japanese Americans. They got around the law preventing release of individual details by showing the relatively few geographical areas where Japanese Americans lived, sometimes down to the city block. In 1942, in the aftermath of Pearl Harbor, that information was used to forcibly relocate and intern over 100,000 people in camps, where most of them stayed for several years.[142]

In 1940, the Comptroller General of the French Army was an enthusiast for punch-card technology named René Carmille. He had visited Dehomag in the 1930s, and in 1934 he proposed a twelve-digit identity number to help with France's mobilisation for the looming war.[143]

In the chaos of the German invasion of France, Carmille took control of as many punch-card machines as he could, of all brands, even ordering new ones. In 1941 he created the new Vichy Government's National Statistics Service in Lyon.[144] Finally, Carmille could create a national identity system that would allocate each person a unique thirteen-digit Personal Identification Number. He proposed a national census, which would, among other things, identify all Jews with a hole punched in column eleven. He also offered to process the specific 'Jewish census' completed in 1941 in both the Vichy-controlled south and the German-occupied north of the country.

After some bureaucratic wrangling, the transfer of 140,000 Jewish census forms to Carmille for tabulation began in 1942. The Reich hoped for an efficient deportation to the death camps like the one under way in Holland, where a national ID card, devised and enthusiastically implemented by demographer Jacobus Lentz, made it easy for the occupiers to pick out their targets, even from a population that actively resisted their efforts. But Carmille's processing of the Jewish census, and of the national census, was not as speedy or efficient as in other occupied countries. In fact, by the beginning of 1944 he still did not have the completed register of Jews that Adolf Eichmann wanted, to fulfil France's quotas for deportation and murder. The Gestapo grew suspicious. In February 1944 Carmille was arrested, along with his office manager Raymond Jaouen. Carmille was tortured by the infamous Klaus Barbie, head of the Gestapo in Lyon, for two days without cease, but revealed nothing that could incriminate any of his fellow officers in the Resistance.[145]

While the Reich hoped Carmille's punch-card machines were readying a harvest of Jewish souls, he was, in reality, preparing information for Allied forces to launch a counter-attack, which began in North Africa in 1942. Local French forces joined American and Allied fighters in Algeria, helped by the information supplied by Carmille's office. Using his position at the heart of National

Statistics, he also prepared false ID papers for many agents and fugitives, even getting equipment and examples to London so identity cards could be made for spies and Resistance fighters entering France.[146]

What Carmille did not prepare, from either the national census or the paper records sent to him from the Jewish census, was a unified register to identify individuals for deportation and murder. Some say he even sabotaged the card-punching machines so they could never punch a hole in column eleven.[147] That register still did not exist when France was liberated in 1945.

Sadly, Carmille did not witness that liberation. Carmille and Jaouen were transported to Dachau concentration camp. Jaouen perished on the journey. Carmille died in January 1945, aged 59.

*

The punch-card systems provided by IBM epitomise how information technology made possible the mass society of the mid-twentieth century. Social security and health systems; employment and production by multinational corporations; mass warfare and organised mass murder – all depended on mechanised data collection and processing that reduced each living person to a number. It is easier to go along with punching code F-6, 'death by special treatment', into a card than with consciously extinguishing the universe of a unique human life. But it is also only possible to mount a war that defeats the Nazis with a detailed and up-to-date knowledge of the people, equipment and supplies you have on your side, where they are and where they are needed.

Punch cards continued to be the standard way to record and

sort data until the 1960s, when magnetic tape began to take over. Post-war computers could handle more information, faster, but still filled entire rooms and required specialised teams to translate human language into computer programs. The Cold War stand-off between two superpowers drove continued innovation in computers for military use, alongside general government administration of mass society, and commercial uses by big business. The possibility of nuclear war provoked a question in American military circles: if Russian nuclear attack destroyed key American centres of intelligence and command, how could the US and its allies coordinate a response?

Scientists at DARPA, the American Defense Advanced Research Projects Agency, came up with an ingenious idea: instead of one network directly connecting the key computing centres, why not build a network of networks with multiple connections between computers? That way, if a nuclear strike took out one centre or network, other connections could continue to send and receive vital information. You can visualise this as the difference between cities connected only by a single motorway, and a network of lanes with many alternative routes. If one road is blocked, traffic can simply take another route to reach the same destination. Nuclear war was not the only scenario the researchers had in mind, however, and they didn't plan for it to be purely military in purpose.

In 1969, the ARPANET first connected Stanford University to UCLA, and other nodes were rapidly added over the next few months. The new technology was demonstrated at a public conference in 1972, and the first email program was introduced in the same year. Because it connected different networks, the ARPANET system was called 'internetting'. There were many technical challenges in developing a system that could, in theory, connect any kind of existing computer network to any other. To reliably send information, it was broken down into smaller 'packets' that would be reassembled, in the right order, at the destination computer. The different networks would also need a common language to send and receive messages.

The two men credited with the invention of the internet are Vint Cerf (also the first to use the word 'internet') and Bob Kahn. Their original working group, founded in 1972, predicted that 256 IP addresses would be enough. At this stage, the internet was seen as mainly a tool for researchers, based in universities, government bodies and corporations.[148] In 1980, 213 computers were connected to the internet, but by 1992, it linked a million machines.[149]

*

In 1965, electronics engineer Gordon Moore wrote an article, 'Cramming More Components onto Integrated Circuits',[150] in which he foresaw that integrating all the electronic components of a computer circuit into a single silicon chip would enable such wonders as home computers and 'personal portable communications equipment. The electronic wristwatch needs only a display to be feasible today.'[151]

Moore played an important role in the invention and development of the microchip, and by the mid-1970s his predictions were coming true. Computers no longer needed to take up a whole room, thanks to the miniaturisation of components. 'Moore's Law', predicting that computing power would get ever smaller and cheaper at a constant rate, was holding true.

Enthusiasts built their own small-scale computers that would fit onto a home desk or table. Steve Wozniak and Steve Jobs presented the prototype Apple-1 computer in 1976. It had 4 kb of RAM and required the user to add a power supply, keyboard, display and storage system. In 1977, it was succeeded by the Apple II, the Commodore PET and Radio Shack's TRS-80, all designed for ordinary people to use without special expertise. Many other companies followed, targeting students, schools and families. IBM's PC (personal computer) was relatively late to the new market, but its software became one of a handful of competing standards.

In 1983, the *New York Times* reported that Britain owned more personal computers per head than any other nation.[152] The UK Government enthusiastically promoted computing, subsidising schools to buy computers and teach programming, and emphasising the educational importance of having a machine at home too. Clive Sinclair, inventor of the Sinclair ZX Spectrum home computer, was knighted for Services to British Industry.[153] The ZX Spectrum sold around a million in its first year, and over five million in all.[154] British Prime Minister Margaret Thatcher presented one to her Japanese counterpart in 1983.[155] The image of every family educating its own children in the new technology that would dominate the coming decades was, of course, a perfect fit for Thatcher's individualistic vision of the future of Britain.

Four more innovations would be needed before we could all start to live our lives in a constant state of connection.

The first of these was the World Wide Web.

Tim Berners-Lee was working at the European Particle Physics Laboratory at CERN on the French/Swiss border, when he began to dream of a computer system that could mirror the human mind's ability to form connections more organically than the rigid mathematical trees and catalogues of machine memory. In 1980 he wrote the Enquire computer program, named after a Victorian book, *Enquire Within Upon Everything*. Now he could make direct links between separate files within one system using hyperlinks. But Berners-Lee wanted to be able to link directly to any file on any system and, just as important, for anyone else in the world to be able to access the connections he was creating.

On 6 August 1991 Tim Berners-Lee published the world's first web page, including information about hypertext and instructions for creating web pages.[156] You can see a copy of that first page at http://info.cern.ch/hypertext/WWW/TheProject.html. In 1994, the World Wide Web Consortium (W3C) was formed, an international organisation to agree standards and systems. Tim Berners-Lee has insisted that the technology underlying the World Wide Web should be freely available to everyone, and refused to patent it or exploit it commercially.

The second step towards perpetual connection was getting everyone online.

When home computers first became popular, the only way to connect to the internet was via the same telephone line used for humans to make telephone calls. Early modems involved putting a telephone handset into a box attached to the computer, which translated the computer's digital signals into audible chirps and whistles. Later models plugged a computer directly to the phone line, but still used audio signals to make contact. The chirruping of a computer 'handshaking' via a dial-up modem will still bring a nostalgic look to anyone who was online in the 1990s. In 1994, one in eight American households had a computer equipped with a modem.[157]

Early mainframe computers had connected directly via telegraph lines, but having a dedicated line for data was expensive. By using a modem, anyone could repurpose their existing telephone line to send and receive data. The transfer rate was slow, and nobody else could make a telephone call while you were downloading a file or checking your email.

The first UK home broadband connection, in March 2000, solved those problems. By 2009 half of UK households had broadband internet.[158] Americans adopted home broadband at a similar rate, going from 1 per cent in 2000 to 60 per cent in 2010.[159]

But before every single home on the planet could get a broadband connection, the third innovation would take over the task of getting everyone online.

In 1973, Bell Labs was working on a telephone system using portable devices, connecting via a network of geographical units or cells. Mobile radio communication had been used by the military since the early twentieth century: Motorola made car radios for police forces from the 1930s, and began to offer a car radio-telephone service in Chicago in 1946,[160] the same year that AT&T started a similar service in St Louis.[161] But the use of these radio telephones was limited by the small number of available radio frequencies, and the large batteries needed for the high-energy radio transmissions.

The cellphone idea would solve these problems by connecting the devices in each cell to a dedicated tower. That meant lower-energy radio connections, requiring less battery power and, crucially, that a few radio frequencies could cover entire countries without interference. Users could move around, knowing that their phone would connect to the nearest tower automatically.[162]

Annoyingly for Joel Engel at Bell Labs, he didn't make the first cellphone call, but he did receive it. Martin Cooper of Motorola phoned Joel from a street in New York in April 1973, to let him know that Motorola had got to a working cellphone first.[163] Early models were large and expensive. Motorola launched the DynaTAC 8000X in 1984, weighing almost a kilogram and costing $3,995. You could talk for half an hour before having to recharge it for ten hours.[164]

America was slow to adopt the cellphone, perhaps because most telephone companies offered free local calls on landlines. In other countries, though, mobile telephones took off fast. Japan's NTT launched a cellphone network covering Tokyo in 1979, but only had car-phones to connect to it. Saudi Arabia got a network in 1981,[165] swiftly followed by Nordic Mobile Telephony, initially covering Norway and Sweden, later Denmark and Finland.[166] The first two UK networks began operation in 1985, and by 1988 half a million phones were connected to a UK mobile network.[167]

By 2002, 60 per cent of Americans[168] and 70 per cent of British households[169] had a mobile phone, but they were still almost exclusively used for voice calls and to send and receive text messages. Meanwhile in Japan, mobile phone users were enjoying i-mode. Launched in 1999, the service used existing handsets to offer email and the World Wide Web. Japan had been relatively slow to get connected to the internet, but now they embraced it through mobile phones. By early 2000, ten million Japanese phone users were connected to i-mode.[170]

In January 2007, journalists and tech enthusiasts queued impatiently to get into Macworld in the Moscone Convention Centre in

San Francisco, some camping overnight to be sure of a seat. The lucky ones were seated shortly after 9 a.m., to see Apple CEO Steve Jobs walk onstage to James Brown's 'I Feel Good'.

After announcing record sales of computers, iPods and iTunes, and introducing Apple TV, Steve Jobs told the launch audience that Apple had already introduced several products that had changed everything. The Macintosh changed the computer industry. The iPod changed the music industry. 'Well, today, we're introducing three revolutionary products of this class,'[171] said Jobs: a wide-screen iPod with touch controls, a revolutionary mobile phone and a breakthrough internet communications device. 'These are not three separate devices. This is one device. And we are calling it iPhone. Today Apple is going to reinvent the phone.'

Steve Jobs was right to talk about the smartphone as something that already existed.[172] Nokia phones running their own software, Symbian, had been available since 2000, offering cameras, apps and a web browser.[173] RIM launched the BlackBerry 5810 in 2002, offering a range of internet services with voice calls almost an afterthought.[174]

The iPhone would run using the same operating system as Apple's computers, Steve Jobs announced. It would synchronise automatically with computers, and its touch screen would use gestures like finger-pinching that are as intuitive today as signing your name with a pen. The iPhone included a camera, accelerometers and a proximity sensor so it knew when you put the phone to your ear for a call, and it used the same web browser as Apple's computers. 'It's the internet in your pocket,' said Jobs.[175]

At the launch, Google CEO Eric Schmidt joined Jobs on stage to talk enthusiastically about working together, with Google providing search and map services via the iPhone. That collaboration wouldn't last long. In 2008, Google unveiled its own smartphone operating system, Android. The Android system was originally developed by a start-up, co-founded by Andy Rubin, which Google acquired in 2005. Instead of trying to manufacture phones, Google made the software available to other companies, so its products would sit by default in the pockets of millions of users, aggregating data.

In North America, smartphones running Android overtook Apple's iPhones in 2010. The rest of the world was slower to shift to the two Californian giants,[176] but by 2013, Nokia announced their last Symbian phone.[177] Today, seven out of ten smartphones worldwide run on Android, just over a quarter are iPhones, and other options are barely a blip on the graph.[178]

In Europe, eight in ten of us use mobile internet services today,[179] and North America is almost as online on the move with 77 per cent.[180] Sub-Saharan Africa, by contrast, has barely three people in ten using mobile internet services, and only half the population has a mobile phone subscription.[181] But if you do have a smartphone – and 53 per cent of the world's population were mobile internet subscribers in 2021[182] – chances are that it's running software written in Silicon Valley by either Apple or Google.

By glorious coincidence, in the middle of writing about the development of the mobile phone, my own mobile phone decided to show me a 1976 clip from the BBC Archives. Two presenters on the popular children's television show *Blue Peter* are demonstrating a curious new piece of technology. Peter Purves takes the show's dog, Shep, puts on a red anorak, and goes out into the Blue Peter Garden carrying what looks like a stylish handbag on a shoulder strap.

'This portable telephone is the invention of an American Scientist, Mr Lew Schnurr,' he tells us, as Shep runs off to investigate something out of shot. Then he extends an aerial from the handbag-thing and dials a number using the rotary dial on the handset. By the magic of television, we see the rotary-dial phone in the studio ring, fellow presenter John Noakes picks it up, and the two have a brief, stilted conversation. Next, it's John's turn to dial; Peter answers with, 'Here we go, every time a winner.' But by now the anorak hood is up. 'I'm getting soaked out here,' he says. 'Do you mind if we don't have a long conversation?'

There's a delicious irony to watching that clip on my own smartphone. But I didn't choose it, or go looking for it. The Instagram algorithm showed it to me because I follow the BBC Archive account. And this is the final piece in the jigsaw of technology that keeps us constantly connected and our data constantly collected – social media.

The earliest users of the web tended to be 'webheads' – people who were technically minded enough to create their own web pages – or organisations who could employ webheads to do it for them. As a more general population got online, the desire to share aspects of your own life with the online world hadn't gone away. Web forums emerged so that anyone could exchange messages without learning to write computer code.

David Bohnett was working for a software company when his long-term partner Rand Schrader finally lost his life to AIDS. This was in 1993, when there were no effective treatments for HIV. Bohnett needed a complete change. He moved house, left his job, and looked for a new project that could absorb him.

The World Wide Web was still new, but it was already gathering momentum within the technology industry. Bohnett decided that what he would build was a new community, in a new space – cyberspace. He recruited John Rezner for his web skills, and the pair founded GeoCities. At the beginning, Bohnett used another new technology, webcams, to draw an audience by connecting them in real time to the real, physical world. He put cameras on street corners in Hollywood so that people could see what was going on, but also go themselves to a bench that was in shot and wave towards the camera, and thus to any of their friends who were watching the live stream. I say 'live stream' – the initial webcam refreshed every eight seconds.

The next step was to invite users to create their own pages, grouped by theme, but named after a city or region. The users who settled this virtual landscape were called 'homesteaders', and by 1997 there were a million of them. Bohnett began GeoCities with no plan for how it would be a viable business, beyond the value of gathering a community who felt some attachment to the service. 'I knew if you built an audience then the business models would follow,' Bohnett told an audience at the Computer History Museum.

When it became a public company in 1998, GeoCities was valued at $86 million. It was the third most visited site on the web in 1999, with forty-one themed neighbourhoods from *Athens and Acropolis* for philosophy to *Baja* for off-roading. The same year, Yahoo! bought it for over $3.5 billion, and Bohnett turned to philanthropy, just before the first dot-com crash pulled the rug from under all the tech projects that had neglected to have a business plan. GeoCities survived the crash, but not the advent of newer social media sites. In 2009, Yahoo! took the digital bulldozers to GeoCities,[183] although the Japanese version of the service lasted until 2019.[184]

Myspace was launched in January 2004 as a rival to earlier social networking site Friendster. Within a month, it had a million users and, in less than a year, five million. But in 2008, Facebook overtook Myspace, which never recovered its dominance.[185,186] Myspace still exists as a site hosting music, video and entertainment news. 'Connect with People,' the site invites you. 'Sign in with your Facebook account to find friends who are already on Myspace!'

As I mentioned earlier, Facebook is now losing ground to video site TikTok. Even parent company Meta's acquisition of Instagram, and its expansion to enable users to post short videos, have not been enough to retain the fickle crowd of social media users.

Some of the short life cycle of social media sites can be explained by technological advance. Today's video-based sites would simply have been impossible when GeoCities began. Neither the hardware that most people were using, nor the internet's connection speeds, were good enough. The short life cycles also reflect the fact that, because we use social media to connect with people, we go where the people are. I think of social media platforms as being like local pubs or cafés, where I go to meet my friends but also to enjoy being surrounded by people I might get to know. Even if I think there's a nicer bar down the road, there's no point in my sitting there alone.

Younger people don't necessarily want to hang out where their parents are discussing slippers and sharing holiday photos, so they're more likely to find new social spaces with their own friends. They also develop new social habits faster. Sending emojis instead of words, then pictures instead of emojis, reflects what's possible, but is also a sign of social groups developing their own language.

The point of social media is for us to interact with each other, either directly with messaging, or via material that we post. We go there to find others like us, to express ourselves to the world, to connect with our real-world friends, family and love interest. We go there to build our identities.

Social media may not originally have been conceived as machines for turning us into data profiles, but if somebody was setting out to design exactly that, it's hard to imagine anything better. In order for us to do anything more than be a passive consumer of what others post, we need a profile. That profile may be anonymous (though social media sites are pushing us to be identifiable, if only to them) but we have a persona which persists in time.

Nick Seaver, who was so perceptive about recommender systems, compares social media platforms with traps. 'Traps are interesting,' he says, 'because they require this projective imagination about prey – you want to understand something about the animal you're trying to trap, you can't just do things with their body, it's also about their minds.' A trap, as he points out, is not just a physical object. It's a form of technology that becomes part of a three-way relationship with the would-be trapper and the intended prey. He makes the obvious parallel with technology that is designed to keep the user returning to it, and reluctant to leave.

In his book, *Computing Taste*, Seaver recounts describing this theory to technology industry researchers. Afterwards, one of them suggested that a better analogy than lethal traps would be the kind of trap made by putting food in the bottom of a jar. When the animal closes its paw around the food, the neck of the jar is too narrow for the fist to leave the same way the extended paw went in. In such a trap, the animal is caught not because it *can't* but because it *won't* let go of the bait.[187]

This is, I think, a very good analogy for how social media is designed. Of course, we *can* leave the site any time we want to – but we don't want to. In a human context, it falls into an interesting grey area between coercion and persuasion, suggests Nick Seaver. 'I think the trap metaphor is useful, because all of these traps require both things at the same time. They don't work unless the entity that you're trying to trap plays the role that's scripted for them.'

Traps also collect data. 'If you put a trap out, and it doesn't catch the animal you're trying to catch, you learn something about how that animal behaves. Contemporary software design is like that,' says Nick. 'People explicitly say, "When we're building software, we're going to make an experimental apparatus that's designed to test market hypotheses to see if we catch users or not. And if we do, to fine tune it to catch them more; if we don't, try to adjust and catch a different set of them." And it's very explicitly that, it's captological.'

Captology – the study of Computers As Persuasive Technology – is a word coined by B. J. Fogg. We'll meet him in a later chapter, when we dig deeper into the more psychological aspects of social media.

But now, I want to move on to the second strand of history that got us where we are today – the history of choice.

Chapter 5: From Necessity to Freedom: Choice

'What do your Levi's® say about you?

. . . Is your longest relationship with our 501® jeans? Or are you a creative type who prefers to use Levi's® as your canvas?

Follow our flowchart to determine which Levi's® style best represents you.'[188]

LEVISTRAUSS.COM BLOG

As a consumer, I have plenty of choice. Some would argue, too much choice. That is part of the appeal of services that offer a tailored menu, whether it's tailored by a computer from my data profile, or by a human who knows me in a more personal way. It saves me from having to look at the millions of options available at the twitch of a finger.

Even now, in the post-Covid economic climate, most of us have far more choice than our grandparents did about how to spend our squeezed budgets. There's a greater variety of stuff on offer and, relative to previous centuries, we can afford to buy more of it.

After the Great Fire of London in 1666 destroyed the stock of the booksellers of St Paul's Churchyard, Samuel Pepys complained that prices went up. Before the fire, he expected to buy a book for eight shillings, an amount equal to a craftsman's wages for five days of labour. After the fire, Pepys paid his bookseller fifty-five shillings for *The Present State of the Ottoman Empire*,[189] over a month's wages for a skilled tradesman.[190] I sincerely hope you haven't paid a month's wages for this book, which shouldn't cost you more than a few hours' worth of minimum wage. Thank you for choosing it, among the thousands of other books on sale.

As well as choices about what to buy, I have choices about where I work and what I do. In the Western world, at least, this freedom is unprecedented. Laws excluding women from particular jobs were still in operation in the mid-twentieth century. Now it's illegal in the UK, and most of Europe and North America, to discriminate against people because of a range of protected characteristics – sex, race, religious beliefs or gender reassignment, for example.

In general, my choices about how to live are far wider than my grandparents had. I can choose what relationships to have, where to live, when and whether to have children, let alone what to wear and how to spend my spare time. In practice, this freedom to choose how to live is bounded by all sorts of practical and social constraints, but any comparison with life a century ago reminds us that a proliferation of choice – or rather, of choices – is one of the characteristics of our age.

How did we get here? Was it simply the inevitable consequence of increasing material plenty? A look at some other political systems across the world suggests otherwise. Not all developed countries are democracies, nor do they all give women equal rights and freedoms with men, or tolerate homosexual relationships. Qatar, which hosted the 2022 World Cup, does not tolerate same-sex or extramarital relationships. At the time of writing, Iran is being wracked by protests against the theocratic regime's strict rules about what women can wear in public. China has combined impressive material progress, lifting millions out of abject poverty into a society of consumer choice, with limiting its citizens' freedom of expression. A comedy club in Beijing was recently fined over $2 million because a comedian made a joke comparing his dog to the People's Liberation Army.[191]

There is no guaranteed link between industrial and technological development and social and political freedoms. But those freedoms would be impossible without technological progress, and the economic growth that first made such progress possible, and then grew to depend on technology. In this chapter, I want to trace the threads of the different kinds of choice we enjoy today, and how they're woven together.

Our journey will end in a tailor shop, but it begins with a haberdasher that we've already met.

*

John Graunt, the self-taught inventor of statistics we met in the previous chapter, was apprenticed to his haberdasher father, and became a haberdasher himself. That meant he, like his father, was a member of the Drapers' Company.[192] Livery companies were like medieval guilds, trading monopolies of skilled craftsmen and their families, which also regulated the training and working lives of their members from apprentice to master craftsman. Companies were welfare societies, networking organisations, and represented their members politically. Graunt's widow applied to the Drapers' Company for a pension after his death.

During the seventeenth century, London was an engine, driving important change on a number of fronts. One of them was the economic shift that was underway, moving from the medieval world of agriculture and craft to a world of finance, international trade, and manufacturing. The old livery companies were being overtaken in power and influence by a new kind of company.

In 1600, Queen Elizabeth I granted a charter to a group of London merchants to form the East India Company, which gave them a trading monopoly for everything east of Africa. The first fleet of fifty ships set sail from Woolwich in 1601, returning from Indonesia a year later with holds full of indigo, silk and spices, and a trade agreement with the Sultan. All luxury goods brought in from Asia had to come through this group of merchants. Investors in the eighth voyage made a 221 per cent profit.[193]

Other companies were formed to trade in other directions. The Royal African Company, founded in 1618 to trade gold and precious materials from Africa, later profited from the transatlantic slave trade, supplying the tobacco plantations of Virginia with forced labour.[194]

Instead of exporting woollen cloth and other home-produced goods, British merchants could now make their fortunes trading the produce of faraway places like China (tea), Indonesia (spices) and the port of Mocha in Yemen (coffee). In many cases, the markets for these goods were initially in other countries, but they soon became desirable commodities in the merchants' home country.

England's first coffee house opened in Oxford in 1651. It was followed the next year by Pasqua Rosée's coffee house in London, where John Graunt met Samuel Pepys, behind the Drapers' Company church, St Michael's Cornhill. Coffee houses were popular places for educated gentlemen to meet and discuss the news, business and politics.[195] Insurance company Lloyd's of London was founded in a coffee house to insure shipping. In this changing economy, the power of the old livery companies declined. Apprentices gained new rights to leave their masters, and even their trades, and then the very institution of apprenticeship was abolished.[196]

The City of London had a huge amount of power. The City funded successive monarchs with substantial loans, which gave it influence over national politics. The monarch granted monopolies of manufacture and trade, enabling the new trading companies to exclude competition and keep prices high. London was also at the heart of political upheavals which first opened the door to the mass of people playing an active role in shaping the future of the nation.

The City had its own militias, first established by Henry VIII in 1537 as the Fraternity or Guild of Artillery of Longbows, Crossbows and Handguns and revived in response to the threat of the Spanish Armada.[197] The Honourable Artillery Company pikemen, still the ceremonial guard of the Lord Mayor of London, in their outfits of red velvet and polished steel, are direct descendants of those 'Trained Bands' of London citizens.

Young John Graunt was a captain in the Trained Bands. As political conflict turned to civil war in 1641, the Trained Bands were reorganised as a disciplined force ready to defend the City, and in November 1642 they defeated the Royalist army at Turnham Green. The City also provided Oliver Cromwell's New Model Army with weapons and money.[198]

In 1647, with his army effectively in control of the country, Cromwell chaired the Putney Debates in St Mary's Church, beside the River Thames. Different political factions within the army argued about the future political direction of the nation.[199] Leveller Colonel Thomas Rainsborough made his often-quoted case for political equality: 'I think that the poorest he that is in England hath a life to live, as the greatest he; and therefore truly . . . every man that is to live under a government ought first by his own consent to put himself under that government.'

This concept, that even the humblest man had the right to consent – or refuse consent – to a government, was very radical at the time. Instead of accepting that there was a natural order, in which the humble folk did what their betters told them, Rainsborough argued that a government only had legitimate power over individuals who had actively chosen to accept its rule: 'I do think that the poorest man in England is not bound in a strict sense to that government that he hath not had a voice to put himself under.'[200]

This call for every man to be treated as a citizen, with the right to withhold consent to be governed, won the vote within the army, though it was ultimately defeated in the country as a whole. But the fact that it was discussed – and voted on – reflects the explosive cultural spirit of the time. England, and especially London, had high levels of literacy and a thriving publishing trade. In a brief period without censorship, ideas were freely expressed and exchanged in public places, first taverns and later coffee houses, in pamphlets and in lively discussion.

Political dissent and argument, scepticism towards authority and a decline of deference were closely connected to religious dissent. Protestant non-conformists like Graunt saw discussion as central to the project of learning about the world as a way to understand God's purposes and live a good life. Freedom to read and argue were central to their vision of a good society. But there were constant political battles over religious tolerance. Graunt himself, after he converted to Roman Catholicism later in life, was taken to court for not attending Anglican church services.

The pilgrims who departed aboard the *Mayflower* in 1620 to found that their colonies in North America were religious dissenters who wanted to be free to live according to their Puritan beliefs. Granted royal permission to settle in Virginia, they had a contract with the Virginia Company, founded in 1606 to colonise the North American coast from present-day Virginia to New York. When they were blown off course, the *Mayflower* pilgrims declared they were no longer bound by the Virginia contract,[201] and instead founded their colony in Massachusetts.[202] Before going ashore, the adult male colonists on board the *Mayflower* made an agreement to 'covenant and combine ourselves into a civil body politic' and to abide by the 'just and equal laws' that body politic would draw up. They anticipated Rainsborough's call that everyone should have a say in the laws that would govern him (and I do mean him).

The Mayflower Compact laid the foundations for the American Constitution, with the idea that all must consent to be governed, and hence must have a say in that government. Again, this in practice meant all European men. But once that principle is established, it's hard to defend it reasonably against the argument that it should include women and non-Europeans too. The Mayflower Compact was the first, very rough, draft of the 'promissory note' of liberty and equality that Martin Luther King would go on to demand the US Government fulfil for all its citizens in 1963.

Seventeenth-century London thus combined three dimensions of choice that would interact for the next 400 years: economic, political and philosophical.

Economic freedom, both the freedom of individuals to move around between jobs, and the freedom of companies to span the world in search of new riches, was the engine for economic growth and for a new model of work that would bring an individual more choice. It also underpinned a new economic system that was founded on colonies, empires, and a new class of people who owned nothing but their own capacity to work. For the next few centuries, therefore, millions of people had less security than under older social systems, and little or no choice about anything much in their lives.

The first stirrings of political freedom briefly opened a door that could not be held shut forever against a mass of people who had glimpsed the chance to run their own lives, and for their choices to have some impact on the society in which they live. We saw in Chapter 3 how that push for political freedom gathered impetus in later centuries.

Freedom of belief, being able to read and discuss clashing ideas, and to decide questions of religion and morality for oneself, opened the door to challenging established beliefs about science, society and human nature. Freedom to choose the principles by which to steer one's own course in the world would eventually become an important moral and political principle.

As a teenager I found it hilarious that 'early modern' history meant going back to the seventeenth century. What was modern about a time when Shakespeare was alive? They still did bearbaiting, wore ruffs, and put people on trial for witchcraft. Now I understand why historians and philosophers talk about that era as the beginning of our own age. The seeds of so much that we take for granted today were planted then.

Amid the whirlwind of the English Civil War, which reshaped the British constitution, discussion of how society ought to be organised had a new urgency. When soldiers debated the future direction of government with their own generals, who were the country's de facto rulers, nobody could justify any political system simply by saying, 'It's always been that way.'

Seeing the civil war coming, English philosopher Thomas Hobbes moved to Paris in 1640, where he wrote his most famous book, *Leviathan*. The title refers to a great and terrifying sea monster, described in the Bible. For Hobbes, the only way men could live peaceably together was to create a shared Leviathan of whom all would be equally afraid. Without that, said Hobbes, men would be constantly afraid of each other, because rational self-interest would lead each to defend himself and his property, and to want what others have. That is the hypothetical 'state of nature' in which, said Hobbes, a man's life would be 'solitary, poor, nasty, brutish and short'.

Leviathan is the first book to talk about a social contract, what Hobbes calls a covenant between subjects that they will all obey the sovereign, or the government. Hobbes defends the institution of monarchy, but not because a king has a God-given right to rule, or because kings are wiser or more just than anyone else. Like Rainsborough, Hobbes claims that the only just source of a king's authority is the freely given consent of the subjects who, in their own self-interest, agree to all give up their absolute liberty, and to all be ruled according to the same laws.

Although Hobbes concludes that a country needs a monarch, or an equivalent ruling body, what makes *Leviathan* revolutionary is the fact that it bases that monarch's authority on the active consent of the subjects. This is the foundation for a society where people can reliably keep their own contracts with each other, knowing that there is a coercive power to enforce them, if necessary.[203]

It also places great importance on the freedom to own property without a constant fear that somebody stronger will take it from you. During the Putney Debates, this fear was used to justify excluding those without property from the right to vote. As Henry Ireton put it, 'no person has a right to this, who does not have a permanent fixed interest in the kingdom. If we take away this law, we shall plainly take away all property and interest that any man has.'[204] In other words, men who owned nothing had nothing to lose, thought Ireton, and might well vote to take away the property of those who did own it, plunging England into Hobbes's state of nature, with each against all, every man for himself. Ever since this time, the right to own property, and to do with that property what you want, has been a central part of European thought on freedom.

That right can come into conflict with other ideas about freedom, though, especially when other human beings are claimed as property. When the United Kingdom finally outlawed slavery, it paid compensation, not to the enslaved people, but to those who had claimed ownership of them. When Europeans settled on other continents, they claimed ownership of land, sometimes by buying it but sometimes by simply taking it, declaring it uninhabited, or saying the earlier occupants were not persons capable of holding legal rights.

In Europe too, most people did not own land or property in the seventeenth century. The old feudal system gave them rights to use a share of the land for specified purposes, for example, grazing animals on the common. But now, landowners were enclosing land and evicting tenants. In 1660, an act of Parliament finally abolished feudal tenures.[205]

An anonymous rhyme of the time expressed popular feelings:

The law locks up the man or woman
Who steals the goose from off the common

But leaves the greater villain loose
Who steals the common from the goose.[206]

In the long run, these changes encouraged innovation in agriculture, and English farming became much more productive and efficient. But in the short run, the enclosures left many people destitute and homeless. Before African people were forcibly shipped to Virginia and the other new colonies to work the fields, English paupers were sent, often as indentured servants with no rights until they had worked a set number of years for a master they did not choose.

Like any time of rapid transformation, the seventeenth century brought brutal change for many people who lost their livelihoods in the paroxysms of economic change. The brief flowering of popular political freedom was extinguished with the restoration of a political order, headed by a king, that did not give a vote to every man. Meanwhile, the luckier people were embracing new social freedoms, liberated from the Puritans' crackdown on secular pleasures like theatres and Christmas, and enjoying exotic new imports like tea and coffee from the East India Company.

This era of international trade and agricultural innovation also sowed the seeds of mass manufacturing. Indian cotton was dyed with indigo, a native plant that produces a deep blue colour. In British homes, imported silk and muslin, calico and chintz rapidly outsold home-grown wool. This angered people working in the wool trade, who rioted in 1697, attacking the East India Company's grand headquarters in Leadenhall Street.[207] But if Indian-made cloth was controversial at home, it found a ready market in China, Africa and the new colonies of the Americas.

The East India Company acted more like a country than a group of merchants. They built their own ships in a purpose-built shipyard on the Isle of Dogs in London. They reached trade agreements with the rulers of other countries. They ran a private army and navy to defend their trading colonies around the world. By the eighteenth century, they were running Bengal, levying taxes and extracting enormous flows of wealth from India to England, while millions of Indians died in famine because of hopeless mismanagement.[208]

In many ways, we still live in a world shaped by the East India Company, and their equivalents in other European countries. It was East India Company tea that was dumped in Boston Harbour during the Boston Tea Party by outraged American colonists, after the Company's newly granted monopoly on shipping tea to America ratcheted up the tax rates.[209] Singapore was originally founded by the East India Company as a trading settlement. The Company's activities in China, trading opium in exchange for tea, led to the First Opium War, which ended with China handing over the island of Hong Kong to the British. The Company was finally liquidated in 1858, its army and responsibilities taken over by the British Government.[210]

There's a deep contradiction in the history that got us from a world of craft guild monopolies and feudal laws to today. Without the complete political, economic and social transformations of the globe, our lives of almost unmanageable choice would never have been possible. But nearly every part of it happened without any choice on the part of most of the people involved, and often against their express wishes.

Companies like the East India Company had no loyalty to the populations around the world whose land, produce and work they took, by trade or by force. They didn't even have any loyalty to the populations in Britain whose work was undercut by cheaper labour abroad. Cotton imported from India, and the new markets for British-made cloth, drove the transformation of how cloth was made, the industrial revolution and the era of mass production.

Britain already had a thriving cloth industry based on home-grown wool, and woollen cloth was among the first exports the East India Company traded abroad. Both the availability of cotton yarn and the expanding markets for cloth, mainly in warmer countries, meant the expansion of what were, initially, home-workers or small workshops. But as demand grew in the eighteenth century, with apparently limitless sources of raw cotton, a series of mechanical innovations radically improved productivity. Arkwright's water-powered spinning machine could turn cotton into yarn hundreds of times faster than a hand-spinner.[211] Power looms came next. The price of cloth plummeted.

At home, undergarments and household linens came within the reach of the lower classes, and in Britain's expanding colonies, products could be adapted to different tastes and spending power. Cheap, British-made cloth was sold everywhere that British ships went, and by the 1800s, that was almost everywhere. The strength of the British Navy, itself reliant on the new production-line manufacturing system, underpinned the network of colonies and international trade that made Britain wealthy.

The Industrial Revolution did far more than produce more goods, more efficiently. It reorganised social life. Instead of working alongside other family members in the same building where they all lived, or on the land around it, individual workers went to work in somebody else's building, by somebody else's clock.

The brutal demands of the machines, which could work

continuously, day and night, without rest, set the inhuman pace of the early factories. As Süssmilch observed in Berlin, factory workers in cities were less healthy than their country cousins. More of their children died. For them, human progress was going backwards. But by bringing a mass of people together in one place, with shared experiences and a common interest in pushing for change, mass production also created the potential for mass political movements.

When Britain's West Indian and American colonies began to grow cotton with enslaved labour, the domestic cotton industry switched to this new source for its raw material. By the mid-nineteenth century, over a billion tons per year were being imported via Liverpool alone.[212] Nearly half a million people worked in the British cotton industry,[213] and while they depended on this slave-grown cotton for their livelihoods, many also felt a sense of solidarity with those who grew it.

In 1861, the American Civil War broke out. Abraham Lincoln's northern fleet blockaded the south, preventing trading ships from bringing cotton to the factories of Lancashire. With no cotton to spin and weave, over half the machines fell silent,[214] throwing hundreds of thousands out of work. The mill owners began to lobby Parliament to send the British Navy to break the blockade. But Manchester's mill workers held a public meeting on New Year's Eve 1862 in support of the blockade. They passed a motion urging Abraham Lincoln to fight on and to abolish slavery.[215]

Their letter to President Lincoln noted the steps already made towards ending slavery, 'exemplifying your belief in the words of your great founders, "All men are created free and equal." . . . It would not become us to dictate any details, but there are broad principles of humanity which must guide you . . . Justice demands for the black, no less than for the white, the protection of law – that his voice must be heard in your courts.'[216] Lincoln wrote back, describing their principled stance as 'an energetic and re-inspiring assurance of the inherent power of truth, and of the ultimate and universal triumph of justice, humanity, and freedom'.[217] In January 1865 US Congress passed the Thirteenth Amendment, prohibiting slavery everywhere in the United States.[218]

American-grown cotton was about to enter a new era, one that embodies the changing shape and texture of choice, through the mass century and into our personalised age.

In 1853, a Bavarian-born wholesaler called Levi Strauss moved to San Francisco.[219] This was four years into the California Gold Rush, and tens of thousands of hopefuls, arriving in California seeking riches, needed bedding and tough work clothes. Strauss provided them to small stores across the west. He also sold cloth. In 1872 a tailor called Jacob Davis, who had been using Strauss's indigo-dyed cotton denim cloth to make waist overalls (hardwearing trousers), with copper rivets to reinforce the stitching, suggested a partnership to manufacture riveted work clothing. Their patent was granted on 20 May 1873, which Levi's celebrates as the birthday of blue jeans.

When the patent ran out in 1890, and other companies were able to copy the original design, marketing became more important. Levi Strauss trousers already had distinctive stitches on the back pocket and a leather patch showing two horses failing to pull a pair of jeans apart. Now the batch number – 501 – was used to identify the original design.

By complete coincidence, this is the variety of Levi's jeans that I wear constantly, either in original indigo-dyed denim or in black.

It was the black pair I was wearing when I disappointed the fashion designer in New York. I was irrationally pleased to discover that I had chosen Levi Strauss's original design. They do fit my body shape, but I daresay I was also influenced by long-ago adverts telling me that this particular cut of trousers is somehow more authentic, rugged, or whichever of my qualities I subconsciously felt my trousers should express.

In the 1920s, America was swept by enthusiasm for cowboys as the embodiment of the independent, frontier spirit. Blue denim became the iconic wardrobe of the western, and of everyone who wanted to be John Wayne, or his Hollywood predecessor Hoot Gibson. Practical clothes aimed at miners and other manual workers were now consumer products that came with an image. Would-be cowgirls could join in too, when Lady Levi's were introduced in 1934.

By now, Levi's had adopted assembly-line techniques to meet demand, in spite of competition from other brands such as Carhartt, Lee – first to introduce zip-fly blue jeans[220] – and Blue Bell (now Wrangler, a brand name that better fits the rodeo rider image). The fact that, until the 1950s, Levi's could only be bought west of the Mississippi added to the Wild West allure.

As the teenage anti-hero emerged in the 1950s, denim was the obvious uniform for the individualist rebelling against respectable society. Marlon Brando wore denim in *The Wild One* in 1953,[221] reflecting its adoption by motorcycle clubs formed by returning servicemen. Some schools banned pupils from wearing denim, instantly adding to its appeal. One 1958 newspaper reported that 90 per cent of American youth was wearing jeans.[222] Civil rights campaigners wore denim jeans. College students adopted the traditional garment of blue-collar workers and farm labourers, not just for practical wear on demonstrations but to make a point: male, female, Black or White, everyone was equal in jeans.

Outside the USA, blue jeans were beginning to symbolise all that was desirable about American life and culture: affluence that also brought more leisure time; youth and physical self-confidence; an independent spirit of self-sufficiency somehow combined with an egalitarian society; and above all, freedom. In 1958, a pair of Levi's went on display in the American pavilion at the World's Fair in Brussels, and the company started exports to Europe the following year. Wrangler launched in Europe in 1962, opening a factory in Belgium.

As the Cold War got chillier, denim jeans became, for citizens of Eastern Bloc countries, both a symbol of freedom and a desirable commodity. My friend Stefan grew up on the Baltic coast of what was, until well into his teens, East Germany. His father was a sailor who brought back denim jeans from the West for his children. Stefan's sister Anja got into trouble when she went to school in American jeans that bore an American flag. She was made to stand in front of the class for a telling-off, but at least she only had to remove the offending flag, not the trousers.

East Germany tried to defuse the allure of forbidden Western denim by manufacturing 'East Jeans' from 1974, but consumers found the quality was not up to the American competition and preferred the illicit trouser market. Many of the Vietnamese diaspora in East Germany worked in the garment trade, and would produce black-market jeans, sometimes with fake labels from Western brands.

Meanwhile, back in America, the role of denim was changing

as youth culture shifted. Mass movements that dressed the same to show they fundamentally *were* the same, not divided by race or sex or class, were fragmenting into subcultures eager to express their differences. Individual self-expression was now a political project in itself. How better to express yourself than by personalising your denim? That way, you could show your allegiance to the tribe of denim-clad rebels while also standing out as an individual.

By 1973, modifying and decorating your mass-produced wardrobe was so popular that Levi's ran a 'Denim Art Contest', judged by an art critic, a photographer, a designer and the curator of San Francisco's De Young Museum, among others.[223] They got thousands of entries, and the winning work went on display in museums and galleries across the country. Levi's jeans had become 'a canvas for personal expression', as the exhibition catalogue put it.[224]

Embroidered, patched, cut and tasselled, some of the garments referenced Native American imagery and techniques. Others included overtly political images about the Vietnam War or Watergate. Many used the fabric as a literal canvas for paint, or reshaped the garment entirely, turning a short jacket into a tailcoat, or trousers into a skirt.[225] Billy Shire won the competition with 'Welfare',[226] a jacket so studded with tacks, rivets and rhinestones that it weighed an extra eleven pounds (Shire ran a studding business). It also featured military insignia, a hotel desk bell and an ashtray.[227]

This kind of self-expression, founded on the affluence that mass production of consumer goods allowed, was explicitly counterposed to the limits of life in the Soviet Union and its satellite countries by Western governments. In America and Europe, the teenagers customising their jeans and jackets were rebelling against those same governments, their parents' values and even capitalism itself. Taking scissors and paint to mass-produced garments was a direct, physical way to tell the world that you despised consumerism: this commodity had value only when an individual used it to express their individuality.

On the other side of the Iron Curtain, denim jeans, and the rebels who wore them, were figures of aspiration. The freedom to be oneself, tearing up convention and defying authority, was also seductive for young people in communist countries, who chafed at the limits of life in authoritarian regimes with failing economies. But for them, capitalism, consumerism and Western politics represented a desirable face of choice that was denied them.

When the Berlin Wall fell in 1989, the citizens of East Germany and other Eastern Bloc countries embraced the freedom to buy consumer goods alongside social and political freedoms. Levi Strauss & Company set up an office in newly re-unified Berlin in 1990, and began selling to a new and eager customer base. But the citizens of Berlin did not tear down the wall, and their own political system, merely to buy denim.

When I went interrailing in newly opened Eastern Bloc countries in 1990, the sense of personal and political possibility was electric. In Romania, a desperately poor country where farmers still used scythes and horse-drawn carts, everyone was eager to share their food with me and to brave the language gap. A year before, they could not have talked to me, a foreigner, without immediately reporting the conversation to the police.

In East Germany, the Stasi's surveillance of every citizen's life was so intensive that any gathering of a dozen people was assumed to include at least one spy or part-time informer. People were sent to prison for putting signs in their apartment windows with innocu-ous slogans like 'When justice is turned into injustice, resistance becomes an obligation!' [228]

The well-founded feeling that somebody is watching and listen-ing has a chilling effect, not just on personal habits and conversa-tion, but on a person's ability to be a full citizen and engage in public and political life. Without the freedom to talk freely in private, it's very hard to develop independent thoughts and ideas, let alone to organise any kind of political movement.

Of course, we should not take the Cold War claims by Western governments to offer freedom and individual rights at face value, nor assume that everyone has freedom of conscience and expression today. European thought may have first developed the idea of the dignity and rights of the individual, and that a government is only legitimate if the people consent to be ruled, but it took a long time and a lot of mass campaigns to turn those ideas into a reality for most in Europe and North America. Worldwide, that fight is far from over.

Even where battles have been won, there is no guarantee that rights and freedoms can't be eroded or taken away by a change of regime, by temporary crisis measures that somehow never go away, or by a changing social climate that values freedom less than safety. That's why we should all campaign for the freedom to do things we never want to do ourselves, on principle. I defend your freedom to go to Mass, or observe Ramadan, or look at pornography, not because I want to do those things myself, but because I want to live in a world where every adult is free to make these choices for themselves.

In this chapter, I have talked about both choice and freedom. They're not the same thing, though they are connected. I sometimes think of freedom in terms of absence of external constraints, and choice as concrete options available here and now. For example, one of the campaigns of the Civil Rights movement was for the freedom of everyone, regardless of race, to sit anywhere at the counter in any diner. That freedom was won by getting rid of segregation. Once people are sitting in a diner, they have a choice of what to order from the menu.

The freedom to take part in public life on an equal footing with other adults is an important political freedom. The freedom to choose from a menu can be a source of great pleasure, but it's not comparable. In fact, it's the kind of freedom we allow children, when we tell them it's dinnertime, and offer them a choice between two kinds of vegetable. The parent sets the parameters of what is available. The child gets to pick broccoli or beans.

We are not children, and our freedom to shape the purpose and direction of our own lives should not be limited to picking options from a menu. We need to be free to do things that are not on a menu, that may come with unforeseen consequences, that might change the course of our lives, or even of the world. In this personalised world, we have choices, but how much freedom do we really have? I will return to this question in later chapters.

*

I was waiting to get on a streetcar on San Francisco's Market Street when the colours caught my eye. I was there in Pride Week so the whole city was fluttering with rainbow flags, updated Pride flags with extra stripes and symbols, and assorted multicoloured celebrations of diversity. But this shop window looked more like a pick-and-mix sweet shop, with jars full of coloured buttons arranged in the familiar rainbow. Inside, I could see more colourful patches and swatches of fabric. The neon sign in the window told me it was the Levi's Tailor Shop, San Francisco.

In a split second, I had abandoned my wait for a tram, and was in through the door. All around me were display cases with attractively laid out ranges of every possible variant on the standard components of a pair of jeans. Leather patch for the back pocket? You can choose from four variants with coloured stripes representing different identities. Classic red pocket tab? You can replace that with a miniature striped flag too. Buttons come in the distinctive stripes or a range of bright and pastel colours – build your own rainbow.

Before you get too cynical about this commercial exploitation of political struggle, Levi's have a long track record of public and practical support for San Francisco's vibrant gay community, going back to the HIV/AIDS epidemic of the 1980s. That was a bold move in a time when homosexuality was being further stigmatised by a new and terrifying disease, and Pride was a defiant refusal to stay hidden, not a marketing jamboree.

It's a sign of how much things have changed here in forty years, that picking a sexual identity off a shelf is now as easy as choosing bootcut or flared jeans. A few display garments, jeans and jackets, gave hints about expressing yourself through denim without needing any sewing skills at all, thanks to the Tailor Shop: self-expression as a service. Part of me found this hilarious: the seventies' spirit of rebellion and individuality, packaged and sold back to the customer. Another part of me was utterly seduced by the idea. As somebody with no sewing skills and no fashion sense, I could nevertheless get a unique garment, by going through a menu and hopefully talking to a nice human who understood clothes better than I did.

Since this was my last day in San Francisco, my trip to the Levi's Tailor Shop had to wait until I got home to London. I got on the next streetcar and began to think about what my personalised jacket should look like, which is where things got difficult, and therefore interesting.

For a start, I didn't plan on adorning myself with any kind of rainbow, or any other stripy flag, patches or buttons. Who I'm attracted to, or in a relationship with, is nobody's business but mine and theirs, and our friends and family when things get serious. I'm not going to wear my heart on my sleeve, my pocket or my bum. But this goes beyond my well-developed sense of privacy. Neither my sexual orientation nor being a woman are central to my sense of who I *am*. When I get mistaken for a man (which happens surprisingly often) or mistaken assumptions are made about my sexuality, I don't feel any less 'me'.

The only time I get annoyed about people – or algorithms – assuming I'm a man is when it reveals a deeper assumption about women in general. If a ticket inspector calls me 'sir', I don't even bother to correct them, but if a motorbike shop emails me as Mr Harkness, I feel aggrieved. 'Well, hang on,' I think. 'Statistically, that may be more probable, but here I am, a real-life woman with a motorcycle, so stop assuming that I don't exist.' Being a woman matters to me when there's some kind of social or political discrimination or stereotyping going on.

So, what about putting something on my jacket that expresses my support for a political cause or campaign? Politics, after all, lives in public. Simon Fanshawe, one of the founders of Stonewall, describes his student life in the 1970s, when 'our political stances were emblazoned on cotton and tin'. He realised that the point was not to stake a claim to virtue with badges and T-shirts, but to live your values openly, acknowledging difference and 'through it finding common cause with others'.[229] One of these common causes was the 'Lesbians and Gays Support the Miners' campaign during the strike of 1984–5, immortalised in the film *Pride*.

I have also worn many political badges and a few T-shirts in my time. Sometimes, I wore them to declare myself part of a tribe. It's especially joyful to be on a demonstration with other people, feeling part of something bigger than myself, visibly belonging to a crowd with one common cause. More recently, as politics has become more fractured and fractious, I sometimes wear a badge that I know will surprise the people I meet, who don't expect me to hold a certain opinion. I wear it deliberately, to challenge their preconceptions and hopefully start a conversation, and possibly change somebody's mind, or at least open it a little wider to different points of view.

I won't be stitching my political ideas and causes into a jacket, though. A symbol or slogan that starts a conversation in one context can too easily, in another, be seen as a declaration of tribal loyalty, or an outright attack on somebody else's identity. I can live my values without turning my clothes into a banner.

How, then, can I use a customised jacket to express who I am? I feel loyalty to the communities my parents came from, though, as a typical unrooted child of the modern era, I didn't grow up in either of them. I want to defend them especially because they're often denigrated and looked down upon. So perhaps a Liver bird and a symbol of Grimsby – a fish? An anchor? The hydraulic tower by the fish dock that was modelled on an Italian Renaissance tower in

Siena?[230] The Liver bird would also symbolise Liverpool FC, which is the only sports team I support, thanks to my dad. I already wear an LFC branded hat, after all.

I have sometimes been tempted to say that I don't have an identity, that I am just a person who does things, but there is an identity that feels important to me. It's not remotely on the same scale as being part of a social group with a shared history of oppression, but it does give me a glimpse of the importance of belonging. It comes with its own language, clothing and bonds of loyalty. It has bonded me instantly with French hoteliers, a washing machine repair man and even the loss adjuster who came round after I was burgled.

There is an instant and direct pleasure in riding a motorbike. It connects you to the road and your surroundings in a way no car can do, feeling the curves in the swing of your own hips, and smelling the trees or the bin lorry or the perfume of the woman crossing the road. But there is also an indirect pleasure in being part of a minority group who share this taste for freedom, this willingness to accept the risk of not being inside a metal box, and a slightly different way of seeing the world. Whether they are in this group or not, I like the fact that other people recognise me as a motorcyclist.

When I was riding in Sicily, a pair of glamorous Italian motorbike police gave me a nickname. I felt pretty chuffed about this and went around telling everyone I was now '*La Tartaruga*', until somebody translated it for me. 'The Tortoise' doesn't sound quite so racy, though it is accurate. Ever since then I have joked about starting a slow riders' club – La Tartaruga MCC. Given the historic association of denim with biker clubs and gangs, that would be a good thing to have on a jacket.

So far, I have got as far as buying the jacket, and am still wondering what I want to put on it. Perhaps by the time you're reading this book I will have decided which aspects of myself I want to wear in public, and emblazoned my sense of who I am on denim. More likely, I will still be unable to pin down exactly what identity I want to stitch into indigo-dyed cotton.

The challenge of using a customised, mass-produced garment to express my identity is the perfect example to end this chapter.

Four hundred years ago, when the Drapers' Company was a political and economic force in London, every garment was handmade, mostly from cloth produced within a few hundred miles. International trade and mass production made possible the era of consumer choice among mass-produced goods, like denim jackets. Seismic changes in social and political organisation eventually brought about the kind of social choices we have today, from the small choices – how to dress – to the much greater ones: what work to do, where and *how* to live.

These social freedoms are the result, not only of vastly expanded productivity, and the replacement of rigid hierarchies with more fluid social relationships, but of new ideas about individual rights. The result has so liberated the individual from social constraints in Western democracies that we can, in a real sense, choose who we want to be. In fact, choosing who we want to be has become a dominant question for each of us, and for society as a whole.

Now we can turn to the most important strand of the past that brought us to the present and is drawing us into the future. The next two chapters will excavate the history of identity, and how it came to shape the way we understand ourselves in the world.

Chapter 6: Pilgrims of Ourselves: Identity

'The Pantheon wants you to be a pilgrim, the protagonist of
a journey that sees you find what you are and what you love,
what you hope for and what you want.'

MONSIGNOR DANIELE MICHELE MICHELETTI, ARCHPRIEST
RECTOR, ON THE PANTHEON'S WEBSITE

The Pantheon is a monument to weathering the storms of history by
running before the winds of change. Its dome, built in the second
century CE from concrete mixed with volcanic stone, is still the
largest ever built from unreinforced concrete or masonry.[231] In
the centre of the hemisphere, a round hole admits a beam of light.
That circle of sunshine, always moving, has traced its slow, curving
path across the dome, and the windowless walls below, many thousands of times. The multi-coloured marble floor is curved to allow
rainwater to drain away. It is a marvel of ancient engineering and
mathematics. Built by the Emperor Hadrian as a temple to all the
gods – the seven main Roman gods – it survived by being converted
into a church in 609 CE. Gloriously unchanged in physical form,
its social role has adapted over the centuries. Today it is rightly one
of the top tourist attractions of Rome.

The church, welcoming all visitors, hopes that some of them will
find themselves on a deeper quest. 'Become a pilgrim of yourself,'
writes Monsignor Micheletti, on the Pantheon's website. 'Seek and

find yourself. Let yourself be looked for and let yourself be found. As for us, we just want to be your travel companions.'[232] Many of my fellow tourists are taking selfies in this very photogenic building. I'm not sure that's what Monsignor Micheletti had in mind. But the whole idea of being a 'pilgrim of yourself' is a transformation of the Church's mission, as radical as the change from Roman temple to Christian church.

Pilgrims have been drawn to Rome for centuries in search of something greater than their own individual lives, greater than humankind, greater than this finite world. Today, says the Archpriest Rector, pilgrims come here hoping to find themselves. It's this transformation of how we understand ourselves, and our place in the world, that I have come to Rome to explore.

Sigmund Freud compared the many-layered human mind to the city of Rome.[233] When he visited, he could still see walls and roads from different eras. Cows and sheep grazed between fragments of columns and arches. The visible city was vastly outnumbered by all the historic cities that preceded it.

'Now let us, by a flight of imagination, suppose that Rome is not a human habitation but a psychical entity with a similarly long and copious past,' he wrote. 'On the Piazza of the Pantheon we should find not only the Pantheon of today, as it was bequeathed to us by Hadrian, but, on the same site, the original edifice erected by Agrippa. And the observer would perhaps only have to change the direction of his glance or his position in order to call up the one view or the other.'[234]

Unlike a city, said Freud, the human mind is capable of keeping the old and the new together. In the mind, a change of perspective is enough to see the equivalents of the ancient temple or the newest church. Unless your mind suffers something catastrophic – the mental equivalent to Mussolini building a road through an archaeological site, as he did with the Via dei Fori Imperiali – all your past is still there. Your first teenage heartbreak is as present as your feelings from this morning, but buried below conscious thought.

I'm thinking about Freud as I wheel my suitcase along the Via dei Fori Imperiali on a hot July afternoon, looking at the seemingly endless historic ruins, stone columns interrupted only by a large cement mixing machine for the construction of a new metro line. Since Freud wrote *Civilization and Its Discontents* in 1930, excavations have revealed, and historians restored, much more of Imperial Rome. The Fori Imperiali – public buildings, baths, temples, and a massive underground sewer – now sprawl across one huge historic site stretching almost to the river. The past here is not so much buried as bursting out in all directions.

I'm here to dig up the history of the self in European thought. On the recommendation of historian Tom Holland, I start with the Basilica of St Clement, first mentioned by St Jerome in 392 CE. Set behind a cloistered courtyard with a fountain, it's a cool refuge from the summer's heat and the city's noise. Through the simple door, hung with white muslin curtains, is a small and gloriously ornate church, glowing with gold and mosaics, the floors tiled with intricate geometric patterns. Except that this is not the basilica mentioned by St Jerome. This church was built in the twelfth century CE, on top of an older basilica, which was partly demolished and then filled in with rubble to provide solid foundations for the new one.

I descend stone steps to the fourth-century brick vaults that run directly beneath the bright, light church. Gentle lighting and smooth walkways take me past fragments of sculpture and surprisingly well-preserved frescoes. Here is the Descent into Limbo, where the crucified but not-yet-risen Christ stretches out a hand of rescue to Adam, treading firmly on a devil that has grabbed Adam's foot. There is the story of St Alexis, who left his family, including his newly wed wife, to go and be holy in Syria. On his return to Rome, realising that nobody recognised him, Alexis lived in his family household anonymously as a servant. His true identity was only discovered after his death. I can't help wondering what emotions ran through his wife and his parents at this discovery.

But this is not the deepest level. I pass the tomb of St Cyril and stand at the top of another stone staircase, deeper, darker and damper than the first. If these levels truly represent descent into the repressed parts of the psyche, Freud might raise an eyebrow at my reluctance. I have never liked small, underground spaces. But I want to see what this one holds. At the foot of the steep, narrow stairs, I peer through an iron grille into a narrow stone vault. Light from a shaft above illuminates an altar, carved with the iconic scene of Mithras, killing a bull while looking up at the sun-deity, Sol.

Beneath the fourth-century basilica lies a first-century Mithraeum. This Mithraic altar lies roughly beneath the altars of both the original and current Basilicas of St Clement. The all-male cult of Mithras was especially popular among Roman soldiers, and some of the key narrative elements may sound familiar – some say Mithras was born in a cave on 25 December,[235] and later ascended to heaven. Rituals involved initiation by purification with water, and shared feasting. The worship of Mithras was suppressed towards the end of the fourth century when Christianity became the official religion of the Roman Empire. I wonder, briefly, how different the world would be if Christianity had not become the official religion of European civilisation, and our society had been built on Mithraic foundations instead. Bad news for bulls, I guess.

By the time I emerge again into the dazzling sunlight, I feel that I've been through an immersive experience of the history of Christianity, and possibly my own subconscious, from the primal elements of wrestling the beast, through stories, images and rituals, and back to the simplicity of a fountain in a courtyard.

The Christian idea of who we are didn't come from nowhere. It drew on Judaism, but also on Greek philosophy and the same Persian religion from which sprang the worship of Mithras.

The idea of humans, unique in all creation, created in God's image, came from the Jewish Torah, which would also become the Old Testament. From the teachings of St Paul came the germ of a universal moral status: that not only Jesus, but every single one of us, is a child of God,[236] that the hand outstretched to Adam in limbo is outstretched to each of us. Over the centuries, the core ideas of Christianity would develop and change, but at its heart this belief, that each and every human is special, would survive.

Around the same time the earlier Basilica of St Clement was being built, Bishop Gregory of Nyssa, in what's now Turkey, condemned slavery for this reason. 'What did you find in existence worth as much as this human nature?' asked Gregory of Nyssa. 'If he is in the likeness of God, who is his buyer? Who is his seller?'[237] In an age when the male head of a household had the power of life and death over wives, children and slaves, a literal patriarchy, this was a radical challenge.

St Augustine, a little later, fudged the slavery question by making a distinction between bodily enslavement and the liberation of the soul through redemption. He also brought to Christian thought a new focus on introspection as the path to God. 'Do not go outward; return within yourself,' he wrote. 'In the inward man dwells truth.'[238]

The history of the Western European idea of the self, like the history of Rome, has both evolved gradually and been wracked by violent change. The ideas of individual dignity and rights protected by laws emerged not just from men (and some women) thinking in quiet monasteries, but from power struggles between popes and kings. The church laid claim to minds and souls, and kings and princes to bodies, and the wealth those bodies produced through work.

In the mid-fourteenth century, a natural catastrophe reshaped European society. The Black Death killed half the population. Entire villages were wiped out in days. Some cities never replaced their lost populations and went into decline. Once the wave of death had abated, the survivors found themselves in a transformed world.

Workers were in demand. In England and France, laws were passed to limit wage growth, contributing to the English Peasants' Revolt, but they could not prevent people moving to find better pay. The feudal institution of serfdom, which tied people to the land where they worked, gradually broke down. In England, half the population, around two million people, were serfs in 1300. By 1500, English serfs numbered just a few thousand.[239] Across Europe, the power of noble landowners declined, and taxes paid in money rose in importance against dues paid in kind, as food or as military service.[240] As Europe put itself back together, Rome had competition from other Italian cities. In Florence the patronage of the Medici banking family provided fertile ground for writers and artists.[241]

Growing respect for reason meant that philosophers were expected to make an argument, not just to cite authorities (human or divine). Writers went back to Greek and Roman philosophy alongside more recent Arabic scholars and early Christian thinkers like Augustine. This rebirth of the classical tradition was first called the Renaissance – *rinascita* – by Giorgio Vasari in his 1550 book, *Lives of the Artists,* and the name has stuck to this day.

Cosimo de' Medici is said to have coined the phrase, 'Every painter paints himself.'[242] Certainly, Renaissance art moved from idealised, almost symbolic representations of figures from mythology or the Bible to figures that resemble their audience. Cosimo commissioned a bronze statue of David, the biblical slayer of Goliath, from the sculptor Donatello. Completed around 1440, it is often cited as the first fully nude statue created since ancient times, though the boy hero is wearing a helmet, crowned with the laurel wreath of victory, and boots. David, slender and youthful, resting one foot on the severed head of his giant adversary, body relaxed as if still catching his breath after the fight, is not a generic youth. He could be a boy who has just stepped in off the streets of Florence.

The renewed interest in the naked human body was not merely sexual: like the Ancient Greeks and Romans, Renaissance sculptors and painters saw in the unadorned flesh something to be revered, not hidden in shame. The naked human form symbolised humanity's central position in the Renaissance world. For those who also felt more earthy desires, there was always Ancient Greece. Marsilio Ficino, another Medici protégé, coined the term 'platonic love' in a commentary on Plato's *Symposium* written for Cosimo's grandson, Lorenzo de' Medici.[243]

One of Ficino's students, Count Giovanni Pico della Mirandola, had already gained a reputation for his learning at the age of twenty-three. Pico della Mirandola studied Philosophy at Padua University, then the most famous in Italy, and at the Sorbonne in Paris. He read not only Greek and Latin philosophers, but also Arabic and Hebrew, going back as far as the Persian sage Zoroaster.[244] The reputation of this witty, educated, young nobleman took a knock when he abducted Margherita, recently married to Giuliano di Mariotto de' Medici, and attempted to take her over the border to Siena. Reports of her willingness differ. Only the intervention of his friend Lorenzo de' Medici, and a large ransom, got Pico della Mirandola out of jail.

The work for which Pico della Mirandola is best known was written as an introduction to 900 philosophical theses, which he planned to debate publicly in Rome. Alas, the Pope did not like his work, the Inquisition got involved, and in December 1487, Pico della Mirandola fled Rome to evade arrest.[245] The planned opening speech was published after his early death as *Oration on the Dignity of Man*.[246]

Mirandola describes the Creator telling Adam, 'We have set thee at the world's centre, that thou mayest from thence more easily observe whatever is in the world. We have made thee neither of heaven nor of earth, neither mortal nor immortal, so that with freedom of choice and with honour, as though the maker and moulder of thyself, thou mayest fashion thyself in whatever shape thou shalt prefer.'[247]

God gave humanity a unique status, telling mankind that 'the nature of all other beings is constrained within the laws We have prescribed for them. But you, constrained by no limits, may determine your nature for yourself, according to your own free will, in whose hands We have placed you.'[248] The word 'humanist' is often used today to mean atheist, but it was first used in Renaissance Italy to describe thinkers like Pico della Mirandola, who put human beings at the centre of philosophy, morality and the arts. For them, it was God who placed man at the pinnacle of creation and gave us freedom to determine our own fate.

Pico della Mirandola returned to Florence in 1488, the year before a talented teenager joined the court of Lorenzo de' Medici to study sculpture. While the humanist writers were putting self-creating man into words, Michelangelo Buonarotti brought forth the human form from marble, bronze or paint. After Lorenzo's death in 1492, Michelangelo worked in Rome for the papal court, before returning to Florence in 1504, to make a statue of David.

Like Donatello's, it was naked. Unlike Donatello's, it was not bronze, and not commissioned by a Medici. Florence was now ruled by a council, elected by a few thousand of the more prominent citizens. It was not a democracy by today's standards, but it was more democratic than life under the Medici dynasty. Among those elected to the council was a Florentine, six years older than Michelangelo, named Niccolò Machiavelli. Today, 'Machiavellian' usually means cunning, ruthless, perhaps amoral. This gives an unfair impression of the man and his writing. Machiavelli was a strong supporter of the kind of Republic that gives a voice to the ordinary people, even if that meant some civic disorder.[249]

Drawing examples from Ancient Greece and Rome as well as contemporary Europe, Machiavelli compared different political systems as well as the characteristics of individual rulers. He did not idealise what people are like, or our capacity to make bad decisions, but he thought the advice of respected men was often enough to turn even an angry mob away from unwise actions.[250] For 'gentlemen' on the other hand – men who live in luxury without working, thanks to their property in land – he had not a good word to say, regarding them as an obstacle to a working republic.[251]

In 1505, while Michelangelo was finding the human form in blocks of marble, and Machiavelli was exploring the practical and theoretical dilemmas of government, a young trainee lawyer got caught in a thunderstorm in what is now Germany. Terrified, he made a vow to St Anna that, if she saved him, he would become a monk. Martin Luther survived the storm and was ordained at his local Augustinian monastery. In 1510, he was chosen to accompany a senior friar to Rome to discuss the plans of Pope Julius II to merge two branches of the Augustinian order. Michelangelo, commissioned by the same pope, was busy painting the magnificent Sistine Chapel ceiling, depicting man as the centre of creation. The opulent reality of the Church's corporate HQ would be a shock to the young, idealistic monk.

After an arduous 850-mile journey on foot, Luther entered Rome at the Porto del Popolo, shouting, 'Hail, Holy Rome!' He made a special trip to ascend the 'Holy Stairs' of the St John Lateran Basilica on his knees, saying the Lord's Prayer on each step. Anyone who managed that complete staircase was said to achieve complete indulgence – being let off the punishments of Purgatory after death – for themselves or a deceased relative.

Luther was disappointed in Rome. If Tripadvisor had existed in 1510, he would have given it one star, accompanied by comments like, 'Rome is a Harlot! Call this the Holy City? Rubbish everywhere, people pissing in the streets, corrupt priests living in luxury, and nobody takes Mass seriously!'[252]

In 1516, the next pope sent monks around Christendom to sell indulgences, raising money to rebuild St Peter's Basilica. In return for a substantial donation, believers got written absolution for sins, reducing the time they would have to spend in Purgatory. For Luther, this represented everything that was bad about the Roman Catholic Church. Instead of being a link between God and human souls, it was a worldly institution to accumulate wealth and exercise power.

Luther wrote in protest to his bishop and archbishop. Pope Leo demanded a retraction, Luther refused and was excommunicated. Put on trial, he again refused to recant, declaring that he was bound by the scriptures. 'My conscience is captive to the Word of God. I cannot and I will not retract anything, since it is neither safe nor right to go against conscience. Here I stand. I cannot do otherwise.'[253] Luther was declared an outlaw, his books banned, a warrant issued for his arrest. But his challenge to the power of the Church and the Holy Roman Emperor Charles V was welcomed by other German rulers, some of whom offered him protection. In hiding, Luther translated the Bible into German, so that anyone could read it without needing a priest to interpret the Latin.[254]

The Reformation that Luther started would, like the Renaissance, put the individual human at the centre of the world. Renaissance writers and artists saw reason and free will as the defining characteristics of human beings, drawing on ancient traditions as well as Christian theology. The Reformation would give a new importance to freedom of conscience, denying that any worldly power could impose beliefs – religious or moral – on a person.

Like so many people in this book, Martin Luther set in motion changes that he didn't foresee and might not have chosen. His emphasis on individual conscience, without the mediation of priests, certainly laid the foundations for an idea of the individual soul that must be free to find its own way to salvation. But, unlike some of his contemporaries in Italy, Luther did not believe that individual human beings could find their path through unfettered reason. It was faith, and following the Bible, that would bring the human soul back to God, who alone could absolve our many sins.

Nor did Luther extend that inward, moral freedom outward into the political sphere. When, in 1524, hundreds of thousands of German peasants revolted against oppressive laws and taxes, Luther had no sympathy with them. Luther's 1525 essay, 'Against the Murdering Thieving Hordes of Peasants', asserts the right of princes and rulers to suppress the rebels with violence.[255] Contrast this with Machiavelli in his *Discourses on Livy* – 'It is easy to learn why this love for free government springs up in people . . . We also need not wonder that the people take awful revenge on those who deprive them of their liberty.'[256]

Machiavelli wrote his best-known work, *The Prince*, in exile after the Medici family returned to power in Florence. He was with the papal army in 1527 when the city of Rome was sacked by the army of Charles V, and died shortly afterwards.[257] Martin Luther died in 1546, after marrying a former nun and fathering six children. Michelangelo lived on until 1564, dying in the same year that Galileo Galilei and William Shakespeare were born.

In their lifetimes, the foundations were laid for some of the changes we have already explored: international trade and money as the driving forces of the new economic order, political and religious dissent, and the overthrowing of old hierarchies. In this chapter, we will pick up another strand of radical change: a strand, driven by ideas, that would transform not only European societies, but how individuals understood themselves.

Citing the maxim that was said to be above the entrance to the Delphic oracle, 'Know Thyself', Pico della Mirandola argued that the way to understand nature and, ultimately, God, is to start by looking inwards. 'For he who knows himself in himself knows all things, as was first written by Zoroaster and later by Plato in the *Alcibiades*.'[258] But the philosophers that followed looked inwards to understand, not God nor nature, but themselves.

*

French aristocrat Michel de Montaigne began a new tradition of introspection in European thought: self-exploration for the purpose of making some sense of one's own inner turmoil. 'I turn my sight inward, I fix it there, I amuse it there,' he wrote. 'Everyone looks in front of him; I look within myself.'[259] In his *Essays*, first published in 1580, he shared, in his words, 'many things that I would not confess to any one in particular . . . and send my best friends to a bookseller's shop, there to inform themselves concerning my most secret thoughts'.

Montaigne didn't think the self he contemplated was a fixed and constant thing, comparing the attempt to 'fix your thought to take its being' to trying to grasp a fistful of water.[260] He also thought it futile to assume that anyone else had the same subjective experience as himself, as we are all unique. His writing is still influential today, but in the century that followed Montaigne, another Frenchman would lay a more substantial foundation for ideas about the self.

If you think of René Descartes, a couple of things probably

spring to your mind. The first one to *my* mind is the Cartesian co-ordinate system in mathematics, which you will have used if you ever plotted the position of a point in relation to a horizontal x axis and a vertical y axis. But it seems I'm unusual in this. More people go straight to Monty Python's Bruces' Philosophers Song in which René Descartes is a 'drunken old fart – I drink therefore I am'.

The other thing most people think of is the original phrase 'I think, therefore I am,' or *'cogito, ergo sum'*, both paraphrases of what Descartes wrote in his *Meditations on First Philosophy* in 1641. He argued that we cannot be sure beyond doubt of anything concerning our bodies, but we can at least be sure that we exist as thinking beings. 'I am, I exist, is necessarily true each time it is expressed by me, or conceived in my mind,' he wrote. 'But I do not yet know with sufficient clearness what I am, though assured that I am.' That split between the material world and the inner world of thought is now called Cartesian dualism.

Unlike Montaigne, Descartes thought that his introspection could be the foundation for a universal understanding of the human mind. His work set the agenda for European thinking about the nature of the human self, and even for the idea that there exists a 'self' that can be studied and described as a distinct substance or object. By putting the thinking mind at the centre, Descartes moved reason from being a property of the universe, or nature, or God, to an activity done by individual human beings.

In a similar way to the Renaissance artists who placed the observer at the focal point of the perspective of a painting, or Luther placing the individual conscience at the centre of the relationship between God and creation, Descartes placed the individual human mind at the centre of philosophy. In so doing, he also turned the human mind into a thing that could itself be contemplated, studied, and even become the object of improvement.

English writer John Locke gave a central role to moral responsibility. In his 1689 *Essay Concerning Human Understanding*,[261] he wrestled with ideas old and new about what makes each of us the distinct, unique, individual we are. Is it the immortal soul, or spirit? Is it the physical continuity of the body? Is it the 'thinking thing' that Descartes described? Perhaps influenced by the Puritan beliefs of his parents, Locke thought the key was a self-awareness that gives us the ability to think and act morally. This self-awareness means each of us cares about what happens to us, individually, and also feels responsible for our own actions, both past and future. Because 'it is by the consciousness it has of its present thoughts and actions that it is a *self to itself* now',[262] this morally responsible, self-directing self is also a reflexive self, looking at itself in the mirror of conscience.

You could regard this as simply a new language for what humans have always done. Instead of the soul, Locke wants us to understand this 'thinking thing'; the thing in each of us that makes my thoughts my own, and your thoughts yours, as a 'self to itself'. But the way we articulate our understanding of what we are shapes how we think about ourselves, and what we do.

Montaigne's fistful of water, a stream of thoughts and feelings, implies that a person's responsibility is to be in the moment, and to direct the flow as best one can. Locke's reflexive self is a thinking thing, responsible not only for what it does, in the past and future as well as now, but what it *is*. We create our future selves through our present thoughts, actions and habits.

Both these frameworks are fundamental to how we understand

ourselves, but they are in tension with each other. How is it possible to be in the moment, and also to be self-aware, to shape oneself for the better? This tension was expressed as a battle within each person between the drive to live authentically as one's true self, and to live respected and esteemed by other people, by Jean-Jacques Rousseau.

For Rousseau, the human sense of self was driven by two different needs. *Amour de soi*, love of oneself, is a natural drive for self-preservation that leads us to take care of our needs for shelter, food, and so on. But in human society we are also driven by *amour propre*, a need for others to respect and esteem us, to find us attractive, to recognise our status. This is a natural drive, but Rousseau worried that in society it can take on damaging forms that lead to everyone playing a part instead of being their authentic self.[263]

In 1761, Rousseau's novel, *Julie, or the New Heloïse*, was a sensation. Countless reprints couldn't meet public demand, and hundreds of readers (mostly women) wrote to Rousseau, many declaring love for an author who saw so clearly into the souls of his characters. Julie, the heroine, is a young woman who falls in love with her tutor, St Preux, but ultimately bows to her father's wishes and marries a more suitable, older man, Wolmar. The novel takes the form of letters between the characters, so instead of an all-seeing narrator, we enter into the characters' inner worlds, expressed in their own words.[264] Does Julie's decision to do what's right represent the triumph of her moral autonomy, or the sacrifice of her true feelings? Is the authentic self the one that can exercise reason and overcome emotion and desire, or the one prepared to flout social convention to follow the inner voice of emotion?

Rousseau lived a life on the fringes of conventional morality. He travelled in France and Italy, working as a musician, a domestic servant and a teacher, and had liaisons with older women, including the widow Mme de Warens, who was his patron for a long time. He lived with, and later married, Thérèse Levasseur, but sent all five of their children to the foundling hospital without even recording their names or dates of birth. This is especially ironic, as one of his books, *Emile*, was a handbook on the ideal way to raise a child.

Rousseau's *Confessions*, published in 1766, far surpass anything Montaigne revealed to his public. 'I want to show a man in all the truth of nature,' he writes at the start of Book 1, 'and that man will be me.' Rousseau describes his sexual awakening as a child being beaten by his foster-sister; his lies, which cost a fellow servant her job; his history of exposing himself to women; masturbation, and much more, inviting his readers to make up their own minds about him. 'I am not made like anyone else I have seen; I dare to believe, not made like anyone else in existence. If I am no better, at least I am different. If nature did well or ill by breaking the mould in which I was cast, can only be judged after reading me.'

It's almost as if Rousseau thinks authenticity and unfiltered self-revelation is enough to redeem him from all the terrible things he is confessing. 'Let the Last Trumpet sound when it will, I will come, this book in my hand, to present myself before the highest judge. I will say loudly, "This is what I have done, what I thought, what I was. I have told the good and the bad with the same frankness . . . I have shown myself as I was; contemptible and vile when I was, good, generous, sublime when I was: I have unveiled my interior just as you have seen yourself."'[265]

Rousseau avoided taking on responsibilities or settling down. He had no stable social role or status until he became a famous author, known not as himself but through the fictional characters he had created. He had no firm moral code, as he switched his religious allegiance but also refused to embrace reason as a guide to doing what's right. Left with nothing but his own inner voice to guide him, Rousseau felt that was the only way to live. Authenticity and transparency, not just being his true self but expressing it to the world, were his defence against losing himself in the clashing social demands and the currents of change through which he moved.

If Rousseau was a restless wanderer, Immanuel Kant was a fixed point, with reason as his guiding star towards Enlightenment. Born in 1724, Kant never left his home city of Königsberg, now Kaliningrad. Inscribed on his gravestone is this quote from his *Critique of Practical Reason*: 'Two things fill the mind with ever new and increasing admiration and reverence the more often and more steadily one reflects on them: the starry heavens above me and the moral law within me.'[266]

Kant was fascinated by astronomy as an example of how human reason can observe the universe and deduce the laws of nature, as we saw in the earlier chapter about technology. But the centre of Kant's moral universe was humanity, embodied in individual human beings. For him, it is our capacity for reason that places us at the centre, because we can – we must – choose how to live and what to do. We must not be treated as mere means towards somebody else's ends. We have the capacity, and thus the responsibility, to set our own ends, our own purposes, and to choose to govern ourselves by a moral law.

Although he expressed belief in God, Kant did not look to the church or scripture as the source for his moral code. Instead, he looked to the reason inside each one of us. It is up to every person to work out the ends or purposes of his or her own life, and the maxims to guide their actions. A life lived by laws imposed from outside can never be a good life, even if all the deeds in it are good deeds.

Kant described Enlightenment not as a condition, but as a process that was taking place as he wrote: humanity emerging from its self-imposed state of ignorance. 'Dare to know!' – *'Sapere Aude'* – was his motto. Political freedom, to Kant, was essential to human progress. He objected to the tendency to describe a population as 'not yet ripe for freedom' because, he said, 'we must be free in order to use our faculties purposively in freedom [and] we never ripen for reason except through our own efforts, which we can make only when we are free.'[267]

At this point, I will note that Kant did not entirely follow through on his own principles. Like many of his time, he left large sections of humanity out of his theoretical idea of humanity. He dismissed women, in the same breath as children, as incapable of reasoning enough for public, political life. Early in his career, he wrote about other races as being naturally different from Europeans, though he did later criticise Europeans who went to other countries under the pretence of civilising them, only to steal and oppress.

Those who accuse Kant of being racist and sexist certainly have a point. But history is full of contradictions. Kant's ideas were central to developing the principles of universal human worth, of equal entitlement to freedom, and of reciprocity as the basis for moral, legal and political frameworks, by which we try to live today.

You could say that the beautiful edifice of thought in which we stand to criticise Kant, Locke and their contemporaries for excluding women, non-Europeans and the lower classes is built over the imperfect, incomplete building that they, and other Enlightenment thinkers, constructed. Like Rome, contemporary Western thought is built from old materials on ancient ground. When we go down the narrow steps to the dark ruins that formed the foundations of newer structures, we can expect to see some ideas that are better buried.

One of Kant's students, Johann Gottfried Herder, did speak out against slavery, and against European cultures who treated other peoples as inferior. He didn't believe that humanity could be divided into distinct races or varieties, insisting that one human species merely came in different local shades. But he was also sceptical of the idea that, through reason, humanity could progress towards one universal, cosmopolitan society.

Herder wrote about *volksgeist* – the spirit of a people, expressed in habits and character traits, but especially in language. He believed that individuals and peoples were shaped by culture. This led him to some frankly racist opinions of the abilities and aptitudes of different peoples, but he defended each people's right to develop undisturbed at its own pace, and to measure progress by its own unique yardstick.

The individual, too, was not to be measured by any universal standard. The particular alchemy of culture and nature within each person could only be expressed by that person, and in expressing that inner nature, the self was given form in the world: not as a rootless, free spirit, but as a particular person within a particular community, speaking a particular language.[268]

This process he called *Bildung*, the individual creating a unique self like an artist crafting a sculpture or painting – *Bild* means picture. The world becomes a mirror to the self, which can only be complete and in harmony with the world when inner nature and outer appearance are at one. Though his name is not well known today, Herder's influence, especially through his philosopher pupil Hegel and his writer friend Goethe, is profound.

The Romantic tradition is often seen as a revolt of emotion against the heartless rule of reason, but the idea of creating one's authentic self like a work of art, reflected in the eyes of others, was just as important. It unites the idea of the self as an authentic inner nature with the project of creating the self by expressing that nature to the world.

English poet and writer Samuel Taylor Coleridge was the first to use the term 'self-realisation'.[269] For Coleridge, self-awareness and deliberate self-formation were as important as listening to the inner voice of nature – one's own nature, as well as nature in general. Both Kant and Goethe had a strong influence on Coleridge, who travelled to Germany to study their work.

Just as the Renaissance thinkers formed their humanist ideas in a time of social upheaval, with military and economic struggles between the Church and secular rulers, Enlightenment thinkers were surrounded by military and political conflicts. The French Revolution was influenced by Enlightenment ideas that challenged the old authorities of church and monarchy. The *Declaration of the Rights of Man and the Citizen*, issued by the French National Assembly in August 1789, begins with the assertion that 'men are born, and remain, free and equal in rights'.

The attempt to make this declaration a political reality inspired both Kant and Coleridge, though both of them felt ambivalent about the bloody reality of a revolution. Kant wrote that the true hope of the revolution lay not in the ongoing power struggles and violent repression, but in the feelings of sympathy the revolution had inspired in onlookers. Coleridge wrote critically of the way

Napoleon later took power, comparing it to the way the Roman emperors took power after the Roman Republic.

While Coleridge was visiting Rome in 1806, Napoleon's French troops occupied Rome's port, Civitavecchia.[270] This was not Napoleon's first invasion of Italy. In 1796 he had occupied Rome and taken the pope to France, where he died three years later. This may have influenced the next pope who, in March 1806, acknowledged Napoleon as 'Emperor of Rome'.

Coleridge was worried that his political writings might make him a target for Napoleon's anger. He left Rome on 18 May 1806, finally reaching the port city of Livorno where he escaped on an American ship before Napoleon's army occupied Rome again in 1808. Through his writings and translations, Coleridge helped introduce the German Romantic tradition to British writers, including John Stuart Mill, who was himself the object of an experiment in the deliberate formation of a person.

James Mill and his friend, the utilitarian philosopher Jeremy Bentham, gave Mill's son John a remarkable education in mathematics, philosophy, history and the classics – he was fluent in Latin and Greek at an age when most children today are playing with finger paints and watching *Paw Patrol*. John went on to have a mental breakdown as an adult, which he later blamed on the fact that he had no independence in forming his own character. Although he was endowed with the best knowledge and habits of his culture, he was given no freedom to decide for himself which parts of his inherited habits and traditions to throw off, and which to keep.[271]

The freedom to find one's own way in life, to be exposed to different ideas and ways of living, and to make up one's own mind about which to adopt, adapt or reject, was central to Mill's idea of a good society. 'Among the works of man, which human life is rightly employed in perfecting and beautifying, the first in importance surely is man himself,' he wrote in *On Liberty*. 'Human nature is not a machine to be built after a model, and set to do exactly the work prescribed for it, but a tree, which requires to grow and develop itself on all sides, according to the tendency of the inward forces which make it a living thing.'[272]

Thinkers in the eighteenth and nineteenth centuries disagreed about what kind of society would be best to allow each individual to fulfil their potential. For Mill, freedom with minimal limits was the obvious setting for individual self-development. For Karl Marx, a capitalist system that turned workers into extensions of the machines they operated, alienating them even from their own work, could never allow them the self-determination necessary to become full human beings. But this way of judging a good or bad society was itself telling. What constituted a good society was now a system that worked for individuals to become themselves. Whereas a good man in Ancient Rome was a man who did what was best for his city and family, a good man in nineteenth-century Europe was one who realised his own potential through self-discovery and self-expression.

John Stuart Mill stood at the crossroads of two ways of thinking about how to be human. The title of his best-known work, *On Liberty*, reflects his concern with human freedom. But he also stood within his father's utilitarian tradition, a school of ethics that drew on the statistical vision of a perfectible society. Quantifying happiness and suffering could lead, believed Bentham and his followers, to rational policies resulting in an increase of total happiness.

In many ways, these two traditions were the two dominant ways of seeing ourselves, as humans, that took us into the mass century. One, claiming the mantle of mathematics and science, seeks to observe, measure and predict ourselves, as well as the rest of nature. The other, claiming the legacy of poets and artists, looks inwards to find an inner nature that each person must find, nurture and realise by expressing it in society. Both these traditions were continued by philosophers into the twentieth century, and we'll come back to some of them in later chapters, but late in the nineteenth century they both emerge in a new way of understanding the human mind.

'It seems to me that perhaps the time has come for psychology to begin to be a science,' wrote William James in 1867.[273] James, who in his youth met John Stuart Mill, would go on to be Harvard's first Professor of Psychology, and write *Principles of Psychology*, which remained the major reference text for over fifty years. Psychology aims to bring the tools and methods of experimental science to our sense of self. One of the most important figures in founding this new science of the human mind is Sigmund Freud, whose comparison of the mind with the ancient city of Rome began this chapter.

In the church of San Pietro in Vincoli is the magnificent marble tomb of Pope Julius II, the pope whom Martin Luther visited, carved by Michelangelo. When Freud was in Rome in 1912, he made daily visits to look at it, and later wrote an essay, 'The Moses of Michelangelo', trying to analyse the emotional state portrayed in the statue of Moses at its centre.

I imagine Freud, the father of psychoanalysis, looking back at Moses through the eyes of Michelangelo. Like the layers below the Basilica of St Clement, the layers of the Western idea of the self are all there: its ancient foundations in Judaism and Christianity, the Renaissance putting man at the centre of creation, and the birth of psychology as a science of the individual mind.

Moses was intended to be one of forty statues on a three-tier tomb, but the commission was cancelled while Michelangelo was working on a set of marble slaves or prisoners – *I Prigioni* – for the tomb's lowest level. Instead, Moses became the centre of a ground-level trio, in a scaled-back, but still magnificent, design. *I Prigioni*, unfinished, are now on display in Florence, their half-formed figures emerging from the marble as if caught in the act of creating themselves.

The prophet, seated and holding two stone tablets, looks troubled and poised as if to get up. Freud, who identified with Moses, was fascinated by the fact that the tablets were blank, with no sign of the Ten Commandments that he was supposed to have brought down the mountain from God. In Freud's time, as in Michelangelo's, Europe was undergoing a crisis of faith in older forms of authority, and Freud was at the vanguard of a new scientific discipline that looked inwards for answers.

Born in 1856 in Freiberg, Moravia, Sigmund Freud was the son of a Jewish wool and cloth trader. The family moved to Vienna when Sigmund was four years old. Inspired by reading a Goethe essay on nature, he decided to study science, and his early work was very influenced by Darwin.[274] When Freud became a neurologist at Vienna General Hospital, he encountered a patient, Anna O., who had hysterical symptoms: partial paralysis, a cough and hallucinations. Initially Freud tried to treat hysteria as a neurological disorder, including with electric shock treatment, then with hypnosis. Eventually he settled on a talking cure without hypnosis.[275]

Freud gradually moved away from treating the human mind as a branch of neurology, and towards an approach that uses each mind's subjective experience of itself, underpinned by a general theory of how human minds develop throughout each person's life.

In 1896, Freud's father died, which affected him deeply. He began to collect antiquities. He told one of his patients, 'The psychoanalyst, like the archaeologist in his excavations, must uncover layer after layer of the patient's psyche, before coming to the deepest, most valuable treasures.'[276] He also began to interpret dreams as a way to access the patient's subconscious mind, including his own. One of the aspects of Freud's own subconscious that he discovered through analysing his dreams was a phobia of visiting Rome, despite an expressed wish to visit the city. In 1897, Freud got as close as the River Tiber, but turned back. He didn't reach Rome until 1901, the first of many visits.

*

It's no coincidence that so many of the people in my history of the European idea of the self have visited, or lived in, Rome. The city has been a driving cultural, as well as political and economic, force, for centuries. The layers of physical relics are not just an analogy for the layers of an individual's subconscious, in Freud's words, but a concrete reminder of the layers of ideas that underlie our modern understanding of who, or perhaps what, we are.

I am visiting Rome, not just to witness these physical relics of European culture, but to meet somebody whose work unites old ideas about the self with new thoughts. I'm here to talk to Paolo Benanti, the author of a book that takes a refreshingly different approach to the question of human identity: 'The person implies identity with oneself, self-possession, which is possible only in freedom,' wrote Benanti. 'This is a dimension inherent to the unrepeatability of every human person, which, called by God as an unmistakable and unique <you>, must also answer personally to the divine invitation.'[277]

I'm drinking blood orange soda at a pavement café next to the Fori Imperiali when Paolo arrives, smiley and apologetic, in shorts and carrying a motorbike jacket and a bag. Just as I arrived, he had texted to say he was running late. He's parked his scooter up the street – impossible to get around Rome any other way, he says – and in the bag is his monk's habit. He opens the anonymous street door and lets us into the cool dark of the corridor. Only a few religious images hint that we're now in a Franciscan monastery.

At Paolo's suggestion we sit outside in the shady garden, birdsong masking the traffic sound. The all-male community are like teenagers, he says, nobody taking responsibility for everyday tasks. In the outside store-room, he indicates a large pack of bottled mineral water sitting beside an empty fridge, to illustrate his point. Paolo has been responsible enough to send the two oldest monks out of the hot city to a cooler place in the hills.

I set up my recording device – two of them, in fact, because I've been caught out before and lost a whole interview. Paolo says a psychiatrist might think I was obsessive, but as an engineer, he's probably a bit obsessive himself. In fact, he's the one who wants to check that the second device is recording because he can't see the waveforms on the little screen.

Franciscan monks work in the community, and Paolo teaches at the Pontifical Gregorian University, founded in 1551. The way humans live with technology – what Paolo calls the 'techno-human condition' – is the focus of his work. As he points out, technology has been embedded in human life for centuries. The glasses he is wearing, for example: 'As a human, I shape the glasses that I wear on my nose, but they shape the world that I perceive. I shape the telescope, and the telescopes shape the universe that I can perceive. I shape a microscope, and the microscope shapes the understanding of human beings. I shape a data centre, and the correlation among data shapes the world, with sense or without sense, to my knowledge.'

A former engineer, Paolo knows a lot about how data can be used to predict what machines will do. 'If I apply this kind of algorithm to a mechanical system,' – for example, a ship in the Pacific – 'and I acquire all the vibrations of the ship, and my algorithms are good enough to predict when the engine will break, that is really useful. I don't lose my ship in the middle of nowhere in the Pacific. I don't stop it for useless maintenance services in the harbour.' This kind of statistical prediction is what data analysis does very well, far better than any human could. When applied to ships, or oil pipelines, or aircraft engines, it puts the power of prediction in the hands of its human operators.

But it can also be plugged into the strand of modern psychology that goes back, not to Freud, but to John B. Watson's 'Behaviorist Manifesto', and to Burrhus Frederic Skinner. B. F. Skinner got his PhD in psychology from Harvard University in 1931 and was a professor there until 1974. Most of his experimental work was with animals, mainly rats and pigeons, but he strongly believed that psychology could both predict and change human behaviour without ever digging into conscious motivations, desires or beliefs, let alone subconscious drives or repressed memories.

To make his experiments on animal behaviour as rigorously scientific as possible, he designed the 'Skinner Box', in which the rat or pigeon could respond to various stimuli by pressing a lever or pecking a key, to be rewarded (or not) with food, or deterred by painful electric shocks[278]. His theory of operant conditioning went beyond simply reinforcing desired behaviour with rewards, or punishing unwanted behaviour. For example, he identified 'superstitions' in pigeons who began to associate random behaviours with obtaining food: 'One bird was conditioned to turn counter-clockwise about the cage, making two or three turns between reinforcements. Another repeatedly thrust its head into one of the upper corners of the cage. A third developed a "tossing" response, as if placing its head beneath an invisible bar and lifting it repeatedly.'[279] In reality, the hungry pigeons were being fed at regular intervals by the automated mechanism. If they happened to be doing the turns or the head movements before the food arrived, and found that after repeating the movement the food arrived again, the pigeon became conditioned to repeat the pointless behaviour.

Another finding was that intermittent rewards were better than predictable ones to keep the animal trying the behaviour that sometimes got the desired result. The rat would press the lever more often when the number of presses needed to get each reward was varied. This insight is used in many human situations, from gambling to app design, because we, too, respond more consistently to inconsistent rewards.

In 1971, Skinner, very pessimistic about the state of the world and especially the behaviour of human beings, published his surprise bestseller, *Beyond Freedom and Dignity*, which argues for a 'technology of behaviour' to solve problems from overpopulation to the 'disaffection and revolt of the young'.[280] The last chapter is entitled, 'What Is Man?' 'It is in the nature of an experimental analysis of human behaviour that it should strip away the functions previously assigned to autonomous man and transfer them one by one to the controlling environment. The analysis leaves less and less for autonomous man to do. But what about man himself?' he asks. 'Is there not something about a person which is more than a living body? Unless something called a self survives, how can we speak of self-knowledge or self-control? To whom is the injunction "Know thyself" addressed?'

The answer Skinner gives could not be further from Gregory of Nyssa, Pico della Mirandola or Coleridge: 'A self is a repertoire of behaviour appropriate to a given set of contingencies.'[281]

This is exactly the kind of reduction of humanity to a system of stimulus and response that Paolo is talking about in the monastery garden. 'If we blur the difference between something that has a cause – a machine – and someone – a man, living beings,' says Paolo, 'this is the real challenge now, because we face machines that are much more like humans, and humans that we understand much more every day like a machine.'

Technology gave us a new metaphor for the behaviourist model of a human being, all inputs and outputs, no need to think about the internal life of the mind, as if Freud had never happened. The apps we use every day are designed to keep us coming back, feeding them with our data, and the companies that use that data depend on our predictable behaviour, as Paolo illustrates.

'So now you are in a church, they are profiling you like someone that sits for an hour in a Catholic place.' Paolo knows this, because Google asked the monks if they could add the location to their maps as a church. 'So, you have a datafication of something that before was not possible to datafy – your spiritual belief.' Except, as Paolo knows, I'm not a Catholic.

'Now, they make mistakes – like us today – but the idea with Big Data is that big is enough, more than accuracy.' In other words, it doesn't matter to Google if the odd atheist gets wrongly put in the 'Roman Catholic' category, if enough correct predictions are made about who might be interested in . . . buying holy relics?

But we, too, want a world that is more predictable. We want some foreknowledge of an uncertain future. Paolo draws a connection with how Romans would have lived thousands of years before. 'You know, there was once a forum here, in which probably someone in love went to the priest in a temple and said, "Should I marry Mary or Joanna?" And he said, "Well, give me a chicken."' The priest sacrifices the chicken, finds in the chicken's entrails some correlations with the heavens that foretell success or failure in marriage, 'And the reply comes from the oracle – "Go with Joanna." Now one day I will ask: "Google: Joanna or Mary?"'

Not Paolo himself, of course – as a Catholic monk, he can't marry Joanna or Mary, but he answers the hypothetical lover in the voice of Google. '"I know you from when you were born. You have a really nice smartwatch – I know all the beating of your heart, with your geolocalisation, because you clicked on 'consent'. And because it's the latest model, I know the impedance of your skin that is connected with the amount of glucose in your blood."' Google will check the questioner's blood sugar, heart rate and skin impedance in the company of Joanna and Mary, compare that to the data of millions of other people, and answer – 'Go with Joanna.'

'We use Google as an oracle,' says Paolo. What's new is not the seeking of guidance for human decisions, but the nature of the guidance we seek through our devices. He calls it invisible. We don't know the inner workings of the digital oracle; the basis for its decision, or even its purposes, apart from keeping us attentive and engaged. Paolo himself was once engaged to be married, and sought advice before making the decision to become a monk instead. But that was spiritual guidance within a long tradition, whose moral principles he knew and shared, and advice from people who know and love him, personally. And Paolo takes responsibility for the decision he made. Without a moral compass, a population nudged by algorithms could have the dangerous illusion of guidance and willingly surrender its freedom and moral responsibility to the opaque workings of a machine. 'It's so deep,' says Paolo, 'the impact of this technology on our understanding of ourselves, that we are reshaping the core question of the human beings.'

Paolo sees this as a change just as profound as the change 'driven by lenses' in Galileo's time. The telescope and microscope changed our perception of our place in the universe: Earth displaced from the centre of creation, and our own world revealed to be teeming with life, too small to be visible to the naked eye. 'We are not, anymore, the centre of the universe, and as Sigmund Freud tells us three hundred years later, that was the first deep wound to our narcissism,' says Paolo. 'We would like to be the centre, [but] we are a part. Now, it's not that infinitely big or infinitely little that we see with lenses, but data captured by a data centre, with correlation, that gives us new tools, changing the way in which we understand the universe. Everything changes – what we are, how we understand – and we have to define again who we are.'

The conclusion of Paolo's book, *Homo Faber: The Techno-Human Condition*, seems to contain two contradictory ideas about who we are. 'Man, a free and responsible being, a spiritual being, a person, experiences his constitution in his deciding and disposing for himself.' So we become fully human, fully ourselves, by setting our life goals and acting towards them. But then, 'He comes to his authentic truth precisely by accepting and sustaining with tranquility this impossibility to dispose of his own reality, impossibility of which he is conscious.'[282] So which is it? Are we free, morally responsible beings who constitute or create ourselves by deciding our own futures? Or do we become our authentic selves by learning to accept that we can't change our own reality?

'I can explain that with Michelangelo,' Paolo says. 'Okay, in Firenze, there are some unfinished sculptures by Michelangelo called [I] Prigioni, like figures that are coming out from the rock. Michelangelo liked to say, "The figure is already inside the marble, I simply take it out." Well, imagine that me and you, when we were born, are like that rock.' Paolo gestures to show the sculptor carefully freeing the figure from the stone. 'Every single free act, if you make it in a coherent way, you have a Michelangelo or Donatello. If you make a . . . ' Paolo gestures wildly with his imaginary chisel . . . 'you have some rocks. This is something that we are, but it's not given before, because we live in history. And so, we have this combination of marble and freedom.'

We create ourselves with our free will, but from the marble in which we are embedded, of which we consist, the substance of our historical time and place, and also of our own individual nature. In

Florence, *I Prigioni* are still caught in that act today, as if struggling to free themselves from inert stone, Michelangelo's chisel marks like a rough sketch of the people they might become. 'We are much more than a machine,' says Paolo. 'We have this deeply inner part in us that is the spiritual one.'

Paolo has to get back to work. As well as teaching at the university he is a working priest in the church around the corner. Before I go, he offers me a coffee from the espresso machine in the little corridor by the dining hall. The religious life would be impossible without coffee, he says. While I drink it, he tells me the history of the monastery. Originally it occupied the whole of this splendid eighteenth-century building adjoining the church. When Napoleon's army annexed Rome in 1798, challenging the power of the Roman Catholic church, the monks were pushed into a few rooms, and never recovered control of the whole building. It's now a hotel. But somehow, the monks continue to live there in a few rooms, most of them underneath the church, 'Like ants,' says Paolo.

Freud would have fun with that mental picture. Deep traditions of belief, a 2,000-year-old moral compass, living on in cracks and foundations, surviving the invasion by post-revolutionary France with its statistically predictable, but hypothetically free and equal, Man. On the roof terrace, the oblivious hotel guests take their evening drinks overlooking the ancient Forum, posting their sunset selfies against the ruins of Roman civilisation.

Chapter 7: The Narcissist in the Hall of Mirrors

'To the performing self, the only reality is the identity he can construct out of materials furnished by advertising and mass culture, themes of popular film and television, and fragments torn from a vast range of cultural traditions . . . In order to polish and perfect the part he has devised for himself, the new Narcissus gazes at his own reflection, not so much in admiration as in unremitting search of flaws, signs of fatigue, decay.'[283]

CHRISTOPHER LASCH

In 1930, the year he was writing about Rome as a metaphor for the many-layered mind, Freud was awarded the Goethe Prize for his work, both psychological and literary. By 1933, Nazis were publicly burning his books in Berlin. Nevertheless, he stayed in Vienna until 1938, when he was persuaded to move with his family to London.[284]

At the end of *Civilization and its Discontents*, Freud tentatively suggests that, if civilisations and individuals develop in similar ways, we could diagnose an entire civilisation, or at least some epochs of it, as 'neurotic'. And, though he puts many caveats and warnings on such a project, he predicts that someone will try to analyse cultural communities as if they were individual psyches. That is, indeed, what happened in the twentieth century. Social theorists began to analyse society, and society's problems, through the lens of whatever psychological theory they thought best explained the human mind. For Skinner, that was behaviourism. Others brought their own ideas.

Theodor Adorno, who fled the Nazis along with other academics in what became known as the Frankfurt School, had begun his career studying Freud's work on the unconscious alongside philosophy.[285] In America, he turned his attention to the urgent work of averting a return of fascism, or its rise in another major democracy. In 1950 Adorno and three colleagues published *The Authoritarian Personality*, which used questionnaires to analyse personality differences. The questionnaire approach was not new. The Myers-Briggs test was already widely used for recruitment, as well as by individuals curious to know more about themselves on four dimensions, with a scientific-looking graph. In fact, Adorno was critical of the classification approach, arguing that 'the desire to construct types was itself indicative of the potentially fascist character'.[286]

He and his fellow authors felt that the root of irrational forces in society could be found in the unconscious urges at work in individuals. Authoritarian family relationships were linked to social and political attitudes that cling to the strong and disdain the weak. At the other end of the scale are 'affectionate, basically equalitarian, and permissive interpersonal relationships', which are implied to correlate with more tolerant and democratic attitudes.[287] Emphasising that personality evolves with social environment, in the family home and wider society, the book takes an almost therapeutic approach to politics. 'For the fascist potential to change, or even to be held in check, there must be an increase in people's capacity to see themselves and to be themselves. This cannot be achieved by the manipulation of people, however well grounded in modern psychology the devices of manipulation might be.'[288]

Alongside political and social bulwarks against fascism, Adorno and his fellow authors conclude, American society and the 'free world' need to ditch psychological repression and embrace free expression of the individual. They must have been cheered by Tom Hayden's 1962 Port Huron Statement, with its emphasis on self-realisation as a political goal.

Erik Erikson was another émigré from Nazi Europe. Trained in psychoanalysis by Sigmund's daughter Anna Freud in Vienna, he expanded the Freudian idea of personality development into identity formation, a process which continues throughout a person's life. He was the first to use the term 'identity crisis' to describe the struggle of an individual to find a coherent sense of self that meshes with their role in society. 'In discussing identity,' he wrote in 1968, 'we cannot separate personal growth and communal change, nor can we separate . . . the identity crisis in individual life and contemporary crises in historical development because the two help to define each other.'[289]

Erikson's own life story reveals a succession of struggles with his own identity. Born Erik Salomonsen, he grew up as Erik Homburger after his mother Karla married Theodor Homburger. Only much later did Erik learn that Theodor was not his father. Karla's first husband, already in America when he was conceived, died not long after Erik's birth, and he never discovered who his biological father was. At a masked ball in Vienna, Erik met Canadian Joan Serson, and they began living together. When she became pregnant, they married, Erik converted from Judaism to Christianity, and in 1933 the family moved to America. Erik and Joan decided to use the Danish custom of using the parent's name as a child's surname, so their son Kai became Kai Erikson. Both parents also adopted the surname Erikson, so Erik became his own father, in name at least.

Erikson described identity as 'all-pervasive and yet so hard to grasp', because the process of identity formation happens '*in the core of the individual* and yet also *in the core of his communal culture*, a process which establishes, in fact, the identity of those two identities'.[290] Erik Erikson worked with Native American children in the Sioux and Yurok peoples, fascinated by the different forms identity formation took in different societies. His 1968 book, *Identity, Youth and Crisis*, captured the mood of disillusionment and disorientation of a generation unable to form a clear sense of who they were in a changing America. Identity moved off the psychiatrist's couch and into the public arena. Since then, it has become an important organising principle in politics, and in society generally, as we saw in Chapter 3. But it also dominates how we, as individuals, relate to the world.

*

I'm on a train towards Amsterdam, feeling very self-conscious about my clothes. I don't usually think about what I wear to visit a museum, especially when I'm travelling alone, in a foreign city, where nobody knows me. But this trip is especially to visit the Youseum, 'the museum of you', and the entire point of the place is to capture your visit in photographs and video, and share it through social media. I'm realising, a bit late, that I generally hate photographs of me and am terrible at taking selfies. Then I start thinking about how I might use the images in relation to this book, and so whatever I decide to wear will be Saying Something about me, and the museum, and this book, and in the end I have opted for a plain black T-shirt and black shorts (because it's very hot that weekend) and the same boxing boots that the fashion designer liked.

On the train, I am reading Christopher Lasch's book, *The Culture of Narcissism*. In it, Lasch did exactly what Freud predicted in 1930, and analysed an entire culture – American culture in the 1970s – in psychological terms. He also pointed out that the culture itself has become more inward-looking, more focused on individual wellbeing than society-wide change: 'The conquest of nature and the search for new frontiers have given way to the search for self-fulfilment,'[291] as he put it.

Thinking in denim terms, it's the transformation of jeans, from the workwear of 1890s gold-rush miners and cowboys, to a canvas for self-expression in the Levi's 1973 Denim Art competition. Lasch described how the narcissistic individual was becoming the norm in American society, not because of rugged individualism, but because of the atrophy of both self-reliance and community systems of mutual aid. Americans had lost both the frontier spirit of courageous risk-taking, and the neighbourly spirit of giving and receiving support.

Because of this brittle sense of self, said Lasch, 'the narcissist depends on others to validate his self-esteem. He cannot live without an admiring audience.' This sounds so much like today's world of social media that I flip back to check the original publication date: 1979. 'For the narcissist, the world is a mirror,' Lasch continued, 'whereas the rugged individualist saw it as an empty wilderness to be shaped to his own design.'[292] Today's narcissist has a screen that not only reflects the individual, but projects their persona outwards to the admiring audience, and feeds back instant validation in the form of likes, comments and shares.

Looking at some of the violent political groups of the late sixties and early seventies, Lasch pointed to the dark side of 'the glamour of modern publicity'. Violent acts are not aimed at seizing political power, or even demanding change, but at attaining celebrity. 'Assassination itself becomes a form of spectacle.'[293] And this, too, is on my mind as the train trundles across the flat lands of northern France, Belgium and the Netherlands. That same weekend, an angry teenager in Buffalo, New York, took guns and audio-visual equipment, and murdered ten strangers in cold blood, as they shopped for groceries, merely for the colour of their skin. The killer live-streamed his atrocity to the internet via a social media site and posted a ranting 'manifesto'. Each of the victims had lives and loved ones. Aaron Salter was the supermarket security guard who died trying to stop the attacker. Katherine Massey was a local activist against illegal guns. Andre Mackniel was buying a birthday cake for his three-year-old son.[294] Pearl Young volunteered at a local food pantry. To the killer, they were not people, but raw material for his fantasy of notoriety and fame. Like many of the news channels, I'm not going to name him.

Even in 1979, Lasch saw the influence of 'the mechanical reproduction of culture' through images and audio recordings. 'We live in a swirl of images and echoes that arrest experience and play it back in slow motion,' he wrote. Smiling for the camera is now an automatic, natural behaviour. 'Cameras and recording machines not only transcribe experience but alter its quality, giving to much of modern life the character of an enormous echo chamber, a hall of mirrors.'[295]

*

The Youseum welcomes visitors like any other small museum, with a cloakroom and ticket desk, but it also provides a guide: not an audio guide, but a human guide. Stan takes me into the start of the tour: two giant paintings, classics of the Dutch masters, but with mirrors where their faces should be. 'Your first challenge is to find your way into the museum,' he tells me. Like an anti-escape room, you have to find your entrance to the puzzle. The solution is more about low cunning than brute force.

We go downstairs and into a little screening room, for a short, upbeat film telling me that I am a work of art, that I deserve to be on a plinth, and that's where the Youseum will put me, says the voice over, 'Gathering dust – only kidding!' Then it's on into the first room, where I literally pose on a plinth among piles of fake money. Stan explains that some of the rooms have cameras to take overhead shots, which I can retrieve using my unique QR code, and print out at the end. He's very patient with my ineptitude as a photographic model.

The whole point of the Youseum is exactly to have fun doing things so you can photograph or video them. I had imagined this as the ultimate in self-centred narcissism, but of course I had it back to front: it's an entirely social experience. No normal person would go alone, as I have done. Most people come in families, in groups of friends, or on dates. The images and videos they make are then shared with a wider circle of friends and family, an extended network of mutual affection and support.

There is a narrative journey. After the money-themed room, and a gold room full of ingots, there is a roped-off backdrop against which to be photographed, red carpet style, and a TV room, which replays you onto the screens of a huge pile of televisions. Then it's on to a warehouse of Likes in boxes, complete with forklift truck, and then a yacht sinking in a sea of Likes. Money? Fame? Social media celebrity? Beware you don't lose yourself. It's a very self-aware museum.

After I've drowned in the sea of internet celebrity, Stan leaves me alone to be reborn – from a velvet swing, through an artfully anatomical vulva, into a hall of mirrors.

This is a surprisingly moving experience. The mirrors are completely disorienting. Some of them are slowly rotating. What's probably a small room appears infinite, with me in my black outfit infinitely reflected in all directions. It reminds me of the end of *The Lady from Shanghai*, the 1947 film noir where the eponymous lady (Rita Hayworth) is shot by her jealous husband in a fairground hall of mirrors, the culmination of a plot in which neither the characters nor the audience ever know what (or who) is real.

After that, the atmosphere gets lighter. There's a green bathtub which you can take either as an environmental statement or an opportunity to pose in the bath. There's an inclusive Chapel of Love, where you can dress up for wedding photos. There's a set with giant teddy bears that will project your message onto the wall: with hindsight I should have projected the title of this book, but it's too late now. Finally, there's a giant ball pit around a cocktail bar called YOUR NAME HERE. At least, that's what the giant letters on its roof spell out.

A birthday party of little girls is having a fantastic time, flinging themselves into the ball pit, swinging over it, pretending to pour cocktails, and generally revelling in the classic childhood mix of fantasy role play and physical exuberance. Cups of juice await them at tables in the little café-and-gift-shop at the end of the tour. I buy a key ring and fridge magnet of the least-terrible photo of me lying on the money plinth. At the time, that seems essential. Later, I regret not buying the YOU ARE A WORK OF ART fridge magnet instead. I'm not sure who would want a key ring or fridge magnet of me laughing at myself on a pile of fake money. The black T-shirt and shorts work, though, so that's something.

The Youseum is the brainchild of Koen Derks, and the day after my visit he invites me into the office behind the museum. It feels young, creative and friendly. 'Before this, I worked as a banker in New York for about four years,' he tells me. 'And in the last year, I was looking for something more entrepreneurial and more creative. I went on a date to an Instagram museum that was just popping up. I paid $45 per person to go in. And it was packed.' Every one of the dozen or so rooms in Dream Machine was different. 'The concept was that people go and take pictures of themselves. I realised, once I was there with my date, there are so many people that if you create a space where they look good, then they will take pictures of themselves. And that makes them feel good.'

Koen wondered what else could be done with this simple idea. 'After the date, I called my cousin Joep, who's my co-founder, and said, "Can you see if this is something that we could do in Amsterdam?" And that was the first step.'

'How did the date go?' I have to ask. 'Was there another date?'

Koen laughs. 'It went really well. I was with her already. And she moved back with me to Amsterdam.' Koen tells me there were similar places in America already, starting with the Museum of Ice Cream in San Francisco in 2015, around the time Instagram first became popular. But the Youseum tries to push the format by including social messages with the fun settings. 'We say we're mission driven, so we have rooms about feminism and climate change and lonely, elderly people, because social messaging is something that's really out there on social media too. We're trying to bring those worlds together: have fun, create content about yourself. And then what we like to do as a team, what excites us, is to also add a bit of a message.'

It's a family operation. 'Everything you see was built by ourselves,' says Koen, 'Jetse and Joep and friends and family around it. I think a good example is . . . ' He calls across the open-plan office to his brother – 'Jetse, do you have the hammock here?' – and then tells me the backstory of the new room they're working on. 'In the Netherlands, there is a scandal about this guy who made twenty million off selling shitty masks to the government. So we're making a room now about this scandal, where you can sit in a big mask. And our mum made the hammock.'

Koen disappears briefly into a storeroom and returns with a hammock-sized medical mask. He looks justly proud of the beautifully crafted prop. The hammock will hang in front of a background of hundreds of masks. It's a collaboration with an investigative journalism company that's just written a book about the case. 'That is spot-on Youseum,' he says. 'People like to sit in something that's a good photo. It's a good story, it's relevant, it's visual. Bang, let's go.'

But it's not just a series of spaces, waiting for the visitor to take a selfie or make a TikTok video. 'We try to take people on a tour that makes them think about their own presence on social media along the way,' says Koen, 'never too heavy and always fun and creative. And then you get a narrative. For example, when you go through the womb that we have, your rebirth in the mirror hall, in the mirror room, that makes sense. But then you're in this reality of social media, where everything you post will always be around you. Because if it's on the internet it's always going to be there. So that's the story we're trying to embed.'

For the next branch of Youseum they're working on a new narrative. 'We are going to train you to be the best "you" on social media that you can. So, you go through a gym where you're trained as a Social Media Superhero, or an activist. All these things related to social media are part of the journey; you check in at the beginning, and you check out to see the difference.'

I clearly need to do that, I tell him. My selfies are terrible. But I suspect I'm not his core market. 'I think our core is two groups,' he tells me. 'First group is families with young children, usually between eight and fourteen. They come as an outing for a day to Amsterdam, and they visit Youseum and they go shopping or have lunch. That's a big audience. And then there's couples or three girlfriends that come and visit together, and they are mostly between eighteen and thirty-five. I think, those two groups are 70 per cent of our visitors.'

For them, taking selfies and posting social media content is just the fabric of life. After my Youseum visit the previous day, I had a beer in central Amsterdam, and noticed that everybody was constantly taking selfies. Even people driving boats were simultaneously taking a selfie. Koen remembers a time before camera phones. He doesn't think the ubiquity of selfies is because people want it more than previous generations. 'I think it's because it's so easy,' he says. 'Because they used to do it. Look at paintings from the seventeenth century. It's just that only rich people had the money to afford a picture of themselves. It's within our veins, it's within our DNA, that you like to see something of yourself and show it to others.'

I remember the classic paintings at the museum entrance, their faces replaced by mirrors. 'And you can argue that, with social media, the picture that people on the whole are showing of themselves is more true,' continues Koen. 'Yeah, of course, there's the Kim Kardashians and so on, but there's also a growing category of people that are using social media as an activist place, or being honest about mental health problems. That was not the case in the 1700s.'

What exactly is it that Koen thinks the Youseum taps into in all of us? 'Narcissism has a bad connotation,' he says, 'but I think there is narcissism in almost everyone. It just shows in different ways. And I think my generation and especially . . . ' – he gestures towards another member of the small team – 'she just turned twenty – [her] generation, I text my friends, she snaps her friends. I think you're very active on Twitter?' I am, yes. 'So that's your portal, or my portal is Instagram,' says Koen. 'It's not that I want to show off that I'm doing really well, it's just keeping in touch with the people that you've collected over the years. That's no different than twenty years ago, it's just a lot easier.'

Every museum and gallery I have visited recently has encouraged people to take and post selfies, with suggested hashtags. Sometimes there is a special location so your selfies look great. All the Youseum has done is cut out the other stuff. We started this excavation of the history of the self in Rome, in the Pantheon, that ancient building that has survived by changing its purpose to suit the changing times. Today, the welcome message on the Pantheon's website is perfectly pitched to our personalised age.

'Become a pilgrim of yourself. Seek and find yourself. Let yourself be looked for and let yourself be found.'[296]

It could be the motto of the Youseum.

*

While I'm in Amsterdam, I pay a quick visit to the Rembrandt House museum. You can see the artist's home and studio, some of his works, and an exhibition of contemporary art. When I visit, there's a display of crying selfies by artist Melanie Bonajo, who describes them as 'a reaction to the prevailing conviction in our society that you always have to be happy, attractive, sexy and young.'

Born in 1607, Rembrandt worked in the Dutch Golden Age, finding a ready market for portraits among the flourishing new merchant class, just as Koen described. Amsterdam was very like London at that time, the early modern age of international trade and colonisation taking shape, alongside religious Reformation and social transformation. Group portraits of the important companies and councils were almost the seventeenth-century equivalent to corporate photographs.

Rembrandt drew, printed and painted dozens of self-portraits. Some of them explore different facial expressions. Some of them are in costume (he kept a store of dressing-up outfits). As a professional artist, they were, in part, a business card to show how he might do your portrait, or your family, or your business associates. Today, he'd be putting them on his Instagram account, but they were also bought as works of art in their own right. The self-portrait as an artistic genre only took off with the Renaissance, as artists became famous individuals instead of anonymous craftsmen, reflecting the rise of the individual human being as a subject of cultural attention. With Rembrandt, it became a genre that expressed the modern artist's quest for self-discovery.

After the Youseum, I can't help asking myself – what *is* the difference between taking a selfie and painting a self-portrait? Is it just, as

Koen says, that now anyone with a smartphone has the technology and the skill required? Is the selfie really just the democratisation of what used to be reserved for the rich and the gifted?

From every one of his many self-portraits, from tousled young man to rueful sixty-three-year-old, in armour or hats or Turkish robes, Rembrandt the model stares back with the same frank look. The intensity of a self-portrait comes, not from the model, but from the artist. Making a drawing or painting takes concentrated looking. It's the stare of the painter looking intently at his model in the mirror that creates the stare of the model, eye to eye with the viewer, on equal terms. When we look at a self-portrait we see, not just the likeness of the artist, but their process of looking and trying to capture something of what they see. It is the artist as subject and object, creator and model.

When I post my inept selfies on social media I'm not showing the result of that kind of intense, focused attention. But I am building up a self-portrait for my online audience. Each photo, which I choose and edit, adds a brush-stroke to my online persona. I am creating myself, engaged in *Bildung*, as Herder or Coleridge might say. But instead of ink or oil paint, I am creating myself in the digital world, using data.

That data is collected and used to create a profile of each one of us, which then becomes the target for personalised adverts, campaigns and services. How that works is the subject of the next chapter. But while you read about the invisible, automated processes that personalise your life, try to remember that this is a two-way process. Each one of us, in our different ways, is constantly creating and editing the identity that represents us in the social world, physical as well as digital.

Chapter 8: Right for You and Only You

'We know what makes you one of a kind.'

EXPERIAN

From the billboard in the station, a goth stares fiercely down at me, hands on hips. She is one of a group of half a dozen similarly dressed people, but the only one looking back at the observer. I remember seeing a video version of the same advert, set in a tube carriage full of them.

'Goths,' began the voice over, before going on to be more specific. 'Goths from the Midlands . . . goths from the Midlands who no longer live at home . . . ' With every narrowing of the category, somebody disappeared from the scene – in one case, leaving a packet of crisps in mid-air. Eventually, there was only one person left who fulfilled the detailed requirements of financial history, credit score and being called Joy ('I thought you said your name was Raven?' protested her last companion, before vanishing). 'Other companies offer similar people similar stuff,' said the voice over. 'At Experian, we use your credit history to help you find deals that are right for you and only you.'[297]

That person staring down at me is Joy, then, awaiting her personalised deals as joyfully as a goth waits for anything except perhaps the end of the world.

This was in 2019, and I remember wondering whether the

advertising campaign (which didn't solely feature goths – I saw another version about anglers) was a response to increasing public awareness of how our data is collected and used to profile us. Most Americans, surveyed in June 2019, were concerned about how their personal data was being used by the government and companies. Four in five were concerned about how much social media sites, advertisers, and companies they buy things from, knew about them. Over two thirds thought that all or most of what they do online is tracked by companies, and four in five thought the risks to them outweighed the benefits.[298]

I don't have equivalent surveys for the UK, but the Pew Research results chime with my experience of talking to people in Britain. Over three quarters of American adults were aware that companies create profiles based on data, and had seen adverts that appeared to be targeting these profiles. Two thirds of those seeing these adverts said that they accurately reflected their actual interests and characteristics.[299]

Here we have the familiar ambivalence about the ubiquitous data collection, profiling and targeting that permeate our digital lives. We don't like feeling out of control, knowing that companies know far more about us than we do about them or how they use our data. But we respond when the algorithms get us right and, as we saw in earlier chapters, we take pleasure not only in getting offers and services that are useful and timely, but in feeling recognised for the individual we like to think we are. Experian was, in a sense, calling our bluff with their advert. Yes, they profile each of us, and yes, that profile is used to offer you stuff that's 'right for you and only you'. Why not? Who wants to be one of the crowd? Part of a tribe, sure, a goth among goths, but still unique and special.

Facebook's parent company, Meta, ran a series of adverts in 2022 that were adverts for adverts – meta-adverts? 'Good ideas deserve to be found' was their theme. Our dissatisfied heroine is idly scrolling on her phone in search of she-knows-not-what, when 'this vegan bakery came sliding down my screen' and she gets very excited. The next guy is bemoaning his lack of cool and his 'saggy, baggy jean' [sic] until an ad for tailoring changes his life.

'Good ideas can be more than stuff we use,' the jaunty voice over continues, 'they have the power to show the world we care about issues.' In this case, by buying eco-sponges. Rainbow-clad dancers, penguins and humans of all shapes, sizes, colours and ages dance together on the street.

Gently poking fun at the randomness of human desires, it's a promise that, however niche your needs and wishes, somebody out there can fulfil them, and it's personalised ads that will find them for you.[300]

The latest buzzword in marketing and 'customer experience' is hyper-personalisation – that is, the promise that every website, product, service and advert will be tailored to each individual. As we'll see in this chapter, that's partly hype aimed at businesses, just as advertising agencies a century ago used psychological buzz-words to dazzle prospective advertisers with science. Nevertheless, personalisation is not a passing fad, it's a deeply embedded trend, though a more honest term would be microsegmentation.

Broad market segmentation has been going on as long as advertising itself. To reach women who buy clothes and cosmetics, advertise in a fashion magazine, not a fishing magazine. To sell cars, put your adverts in the motoring section, not the children's section. Microsegmentation makes it possible to reach very niche audiences, down to individuals, but even individuals have been profiled since the very beginnings of the mass market. Today's data brokers, like Experian and Acxiom, have their roots in credit reference agencies and direct mailing lists.

Experian's company history identifies the first ever credit reference agency as a group of London tailors who, in 1803, started pooling information about their customers. By 1826, Manchester innkeepers and other traders had formed the 'Society of Guardians for the Protection of Tradesmen against Swindlers, Sharpers and other Fraudulent Persons', which circulated a monthly newsletter listing people who had not paid their debts.[301] In America too, the emerging mass market was fuelled by credit, but people could easily escape their debts by moving to another city or even another state. Old systems of trust, based on knowing people in their family, community and workplace context, didn't work in this new world of mobile individuals.

One man who felt this sharply was New York silk wholesaler Lewis Tappan. Bankrupted by unpaid debts that he could not collect, he started a new business, intended to protect others from a similar fate. The Mercantile Agency, founded by Tappan in 1841, compiled a database of business owners across the United States.[302] Within thirty years, Tappan had over 10,000 local agents, actively investigating the character and reputation of each business owner and the state of the business, and centralising that information to the agency's vast library of ledger books in New York. Initially, subscribers sent a person to put specific queries to a clerk, who read the relevant ledgers and gave a verbal summary. Later, the agency's reference books were encoded for those who paid for access. It went on to become Dun & Bradstreet, still one of the world's leading credit rating companies.[303]

As the mass market expanded in the twentieth century, credit agencies became more important, and computerisation of records in the 1970s made it much easier to combine data sources on individuals. Great Universal Stores sold goods on credit to around a quarter of the UK population, so their database was a de facto credit record of their customers. In 1980, that formed the basis of a new venture, Commercial Credit Nottingham (CCN),[304] which offered automated credit scoring and innovative networked systems that let clients instantly check credit scores. In 1985, CCN started to use the information it held for direct marketing.[305]

Direct mailing lists were almost as old as credit agencies. In America, especially, where customers might be spread across great distances, mail order was an important sales channel. The Sears Roebuck catalogue in 1895 offered 532 pages of assorted goods from furniture to firearms, hats to horse-drawn carriages.[306] One especially lucrative product line was mail-order medicine.

When Louen V. Atkins was shot dead by his former business partner James Rainey in 1910, after a dispute over a medical mailing list escalated into violence, Rainey's trial revealed the scale of the letter-broking business.[307] Newspaper adverts encouraged people to write in, listing their symptoms and other information, including name and address. These letters were used to sell directly to the poor sufferer, and also sold on to agencies like the Medical Mailing List Company of New York, which rented out letters in batches of a thousand. You could order letters by type of ailment, by rural or city, and by how recently the letter was sent.[308]

Ralph Lane Polk started out as a door-to-door salesman of patent medicines, before switching to compiling and printing directories in 1870. R. L. Polk & Company amassed data on millions of individuals and households, and when Ralph Lane Polk II took over the business in 1923, he started a Direct Mail Division. Ralph Lane Polk III, taking over in the 1950s, computerised the databases,[309] which by now contained over sixty million names.[310] Sadly, Ralph Lane Polk IV only served four years as company president before disappearing in a boating accident in 1985,[311] but the company continued to expand, acquiring both a consumer database company that used questionnaires to demographically profile millions of names, and a digital mapping company.[312] It's now part of credit ratings agency S&P Global.

When computer-readable census information became available for research, social scientists were the first to dive into the rich ocean of data to profile people by where they lived. In Britain, Richard Webber used the 1971 census to map inner-city deprivation in Liverpool, effectively inventing geodemographics. Webber's Classification of Residential Neighbourhoods project went on to be the basis for the ACORN classification system, now widely used to describe the populations of areas as small as a single postcode. Without knowing anything more than an individual's name and address, characteristics such as ethnicity, income and consumption habits can be predicted well enough for public policy decisions, as well as marketing.

Claritas, established in 1971, used the United States census to classify Americans into forty groups, with 'hard scrabble' at the bottom of the social ladder, and 'Blue Blood Estates' at the top. Adding in ZIP codes tied these classifications to small neighbourhoods. Claritas claims to use over 10,000 variables, including vehicle registrations and consumer surveys,[313] media usage information from Nielsen, and shopping history.[314] Their Canadian wing offers to classify any Canadian postcode into one of sixty-seven segments. 'Vie de Rêve' is French-speaking Canadians 'living the dream' with an affluent suburban lifestyle that includes skiing, comedy shows and 'wine-and-whatever parties'. The largest segment, 'Down to Earth', represents one in forty Canadians, mostly older families living in rural areas, who like sewing, birdwatching, country music radio and snowmobiling.[315]

These profiles, characterising a neighbourhood socially as well as demographically, are the product of computers using statistical methods on huge quantities of data. Human beings find it difficult to imagine a graph that has, not the two x and y axes that Descartes invented, or even three axes in three dimensions, but thousands of axes, each one measuring a different dimension like 'number of snowmobiles per household' or 'amount spent on wine per year'. Computers have no such problem. Using a technique that goes back to Victorian times, computers can find clusters of data points that are close together, in this imaginary thousand-dimensional space, and overlap very little with other groups. Then, the computer combines the most predictive characteristics of these groups into a few values, so humans can grasp what makes cluster twenty-three different from cluster twenty-four.

Richard Webber has described these clusters as 'natural' because, instead of the human researcher imposing a theoretical framework on the population being studied, the groupings emerge from existing data, and the labels are applied later to the results of the observation.[316] Seen this way, the classification is more like a naturalist trying to distinguish a species of moth from other moths, and then name it, than a physicist predicting the existence of a particle and then devising experiments to find out if it exists. The test of this geodemographic method is whether the classified groups do indeed behave differently from one another. For marketing purposes, they predict quite well what people will buy, watch, eat, and where they like to travel. They also predict voting behaviour fairly well; certainly better than older measures of occupation and income alone. But before any kind of computer can use your profile to target you, it has to do one thing that is much harder for a machine than for a human being – it has to know who you are.

I recently had a baffling phone conversation with Allan, a lovely motorcycle instructor I had not spoken to for ten years. He rang me because I had just texted him to say I had replied to his email. It was baffling because neither of us knew who the other person was. I thought I knew who *he* was. He was Allan. My phone said so. Allan, the person I had texted many times, and whose email I had just answered with a text saying, 'Any chance we could do it earlier?' But even when I started to realise that he had no clue who I was, why I had texted him, or what I was talking about, and I asked who he was, of course he said, 'Allan,' because that's his name. By now, I had noticed that he didn't have the Scottish accent I was expecting. So I had to ask, 'Which Allan?'

It's not an easy question to answer, out of context. If you're at a wedding and somebody says, 'Which Allan?' you can say you're the bride's cousin or the groom's mother's second husband's nephew. If you're in a social group and know who the other Allan is, you can say, 'Not Allan with the beard, Allan who plays the viola.' But if you don't know who's asking, what do you say? How do you differentiate yourself from a completely different person who is not you but happens to have your name? Luckily, he replied, 'Allan the motorcycle instructor,' and I remembered him, and asked how business was, and then he went off to walk his dog, and it was all fine.

How did I get into this mess? My phone, trying to be helpful, found two Allans in my contacts – Allan the motorcycle instructor and Allan Grant, an agent sorting out some work for me, and decided they must be the same person. So my text to Allan Grant went instead to Allan the motorcycle instructor, and I am very glad that it wasn't anything embarrassing.

This is the most basic level of identity for a person – distinguishing ourselves from other people. When I have to show ID to collect a parcel or board a plane, I am offering proof that I am precisely the Timandra Harkness to whom the parcel was addressed, or whose passport I am holding, and not any of the other Timandra Harknesses out there, if there are any. Each of us, on that level, knows who we are. The teacher may get two Catherines in the class mixed up, but every afternoon, each of them knows her own way home without stopping to ask, 'Now, which Catherine am I?' This utterly common-sense idea, that I am me and not you, is surprisingly hard to translate into the language of machines.

In formal logic, the equals sign is used to say, 'This thing and that thing are the same thing.' I remember being driven crazy in a mathematical logic course where I kept losing marks for not writing 'x = x'. It made no sense to me. If x *doesn't* equal x, then what does? But it becomes more useful when applied to real things or real people. If we want to know whether the statement 'Socrates was a philosopher' is true, we need to know whether the 'Socrates' in that sentence is the man who lived in ancient Athens, or the adorable-but-dim rat called Socrates I used to have as a pet. We might have to specify that 'Socrates' in this context means 'the man who lived in ancient Athens, was married to Xanthippe, taught Plato and Xenophon, and was executed by drinking hemlock'. *That* Socrates was a philosopher.

Socrates the rat didn't understand these subtleties. I remember him sitting on my knee one evening, getting very excited because Bettany Hughes was on the television and kept saying his name. She was talking about Socrates the Greek philosopher, obviously, but Socrates the rat didn't know that 'Socrates' could stand for anything except him. He wasn't very smart, even for a rat, but I was fond of him.

Apparently, he was as smart as my smartphone, though, which couldn't tell that two Allans are distinct people who just happen to share a first name. If a computer is trying to identify you, whether that's to let you take money out of your online bank account or to target you with advertising, that could be a problem. They might end up letting one Allan take out money that the other Allan had earned doing motorcycle training. They might target Socrates the rat with adverts for wine and philosophy textbooks, or Socrates the philosopher with adverts for a new hammock, because you have chewed through your old hammock *while lying in it*, landing you with a bump on the floor of your cage *for the third time*. As I said, he wasn't very bright. The rat, not the philosopher.

Identifying people we know is a basic human skill, so taken for granted that people who suffer from prosopagnosia – face blindness – struggle in a society that takes not being recognised as an insult. But it's an inter-human skill that computers don't always handle well. Bethany K. Farber is suing the Los Angeles Police

Department for $2.5 million after spending nearly two weeks in prison because her name appeared on an arrest warrant in Texas. Bethany has never been to Texas. She was trying to board a plane for Mexico when airport officials stopped her and called the police. Refusing to check her date of birth, middle name or passport, the police handcuffed her and kept her in prison until her lawyers were able to show she had not been in Texas when the crime was committed, using location data from her mobile phone.[317]

Suppose you're a data broker, and you want to know whether I am the same Timandra Harkness that bought three rat hammocks online, to decide whether I fall into the category 'owners of rats who never learn'. You don't want to ask me, and if you did, I probably wouldn't answer truthfully. Distinguishing between people who have the same name becomes easier for computers, the more data they have on each individual. Generally, it takes only a handful of data points to tell two people apart. Even if they're both called Ralph Lane Polk, and live at the same address, they should have different dates of birth.

Let's say I arrive at a web page, or on a social media site, linked to an ad network. If I have logged in to a profile on a social media platform or commercial site, that's also my profile for advertisers. But if I don't log in, I can still be linked to a profile using cookies, or the device or IP address I'm using, or even my browser fingerprint. To find out how easy this is, go to the website coveryourtracks. eff.org, and the Electronic Frontier Foundation will tell you how special you are. Today, my browser fingerprint is unique among the 191,021 tested in the last forty-five days. That makes it very easy for the websites I'm visiting to recognise, not me perhaps, but my machine and the software I'm using.

Every time my computer takes me to a new site, it automatically transmits innocuous information to the website I'm visiting, like what browser I'm using, time zone, fonts on my system and available memory. None of these things is especially revealing but – like the unique combination of age, postcode and household size – together, they swiftly narrow my computer down to a category of one. So, even though I default to saying no to cookies, and turn my privacy settings up so high that some websites refuse to work at all, my computer is still as recognisable to other machines as my car is to any Automatic Number Plate Recognition system.

Using a different device won't necessarily help you stay anonymous either. Spotify offers a running playlist that selects tracks to match or even increase your pace. It can do that because your mobile phone detects your running tempo, which is very distinctively yours. Behavioural biometrics collects data from what people do, our gait or how we interact with a device. While we scroll through social media or bash out texts, the phone or tablet is recording the idiosyncratic way we type or how long we take to release one key and hit the next one.

Companies like TypingDNA[318] use AI to analyse those patterns, offering what they call frictionless authentication. They insist that their clients must tell users that their typing is being used to identify them, but national security agencies would have no such qualms. They might find it very important to know that person x logging in to encrypted systems from a public computer is the same person x who ordered a takeaway from a home computer using an identifiable credit card. In other words, that $x = x$.

There is a technological and legal arms race between campaigners trying to protect our privacy, and companies whose entire business model is based on targeted advertising. Governments sit somewhere in the middle. Many governments are trying to defend the rights of their citizens, and to provide a counterweight in the power relationship between individuals and tech companies. Europe and some American states are gradually introducing legislation designed to make individuals more aware, at least, of what data is being collected about them. But governments, too, want to be able to identify us, especially as more and more government departments relate to citizens remotely.

Attitudes vary strongly between different countries. In Sweden, every resident needs a Personnummer, a twelve-digit identification number, to do all sorts of basic things like open a bank account or get healthcare.[319] This is generally accepted in Sweden as being the best system for a smooth-running modern society in which the state provides lots of vital services. In India, the Aadhaar system uses fingerprints and a photograph, but is not compulsory and is still regarded with suspicion by civil liberties groups. In the UK, repeated attempts to introduce a national ID card system have been met with resistance from campaigners who don't feel that individuals should have to prove who they are on request.

Nevertheless, the introduction in 2023 of a requirement to show photographic ID to vote in the UK means that there is a kind of ad hoc national ID system for anyone who wants a democratic voice. That has accelerated the growth of privately run identity card services, providing people who don't have a passport or driving licence with something to prove who they are. One example is CitizenCard, founded in 1999 and run by a consortium including Experian and Yoti, a digital ID company that uses biometric data (a face scan) to link your profile with you. Yoti's selling point is that, unlike showing your driving licence to prove you are old enough to, for example, buy alcohol, showing the app only shares your face and the simple confirmation 'OVER 18', not your address or exact date of birth. Even if some governments refrain from imposing a national digital ID system, I predict that every adult will soon have some kind of digital profile that uses their face to link their physical with their online existence.

Let's assume, then, that the website or app you've clicked on has linked you to a profile. How does it know what adverts to show you? This technology is not as new as you might think. A journalist called Melanie Warner, writing for *Forbes* magazine in 1996, described the now-familiar experience. 'You sign on to your favourite website and voila! – up pops an ad for Happy Times Cruise Lines. "Escape the Northeast winter blues with Happy Times," beckons the ad. "Click here to find out about our exciting executive promotion." Sure enough, you work in Connecticut, and you've been thinking about vacationing in the Mediterranean. But how do they know that? Whoever they are.' Nearly thirty years ago, when the web was young enough to have a special word for people who used the web, Warner described the birth of micro-targeted advertising, and identified its parents. 'The omniscient "they" is probably DoubleClick, an advertising broker that serves both companies with websites and advertisers looking to reach Webheads.'

DoubleClick, founded in March that year, had built profiles for four million people within a few months, including their email address and web pages they had looked at, and made lists of their interests in categories like gardening or sports. As Warner describes it, every time you log on to a DoubleClick site, the software identifies you, 'checks out your user profile, and uploads an ad customized for you – within milliseconds of your signing on'.[320] Google bought DoubleClick in 2008, integrating it into its own systems, and quickly became the world's biggest internet advertising company. Advertising is the main source of revenue for Google and its parent company, Alphabet, which made almost $150 billion from Google ads in 2020. These ads appear with Google search results, of course, but also on YouTube, Google Maps, and other Alphabet-owned services, as well as on other companies' websites.[321]

The principle underlying this business, which makes Alphabet one of the most valuable handful of companies in the world, remains the same as Melanie Warner's advert for cruising holidays. It's about displaying adverts that are 'right for you and only you,' in Experian's words. The process that chooses the perfect advert for you, on that day, on that website, at that time, is called Real Time Bidding, or RTB.

Real Time Bidding is literally that: in the fractions of a second it takes a web page to load, an app to open on your phone, or your search results to arrive, automated auction systems are comparing bids from advertisers for the opportunity to show you an advert. They're not just bidding for five seconds of video time, or for a certain patch of your screen. They're not even just bidding for time and relevance, the way TV advertisers pay more to appear with popular programmes, or newspaper advertisers pay more where the editorial content suits the ad. They can set different bids for mobile devices, dates and times, locations, but they're also bidding for the attention of the right audience.

Google Ads Help describes in detail how you can weight different aspects of that audience by adjusting the bid amount. In their example, you might bid $1 for a 'soccer fans' affinity audience, add

20 per cent for a 'male' demographic, and another 10 per cent for the geographic specification 'in Argentina'. So, you'd pay $1.32 if a male soccer fan in Argentina sees your advert, but only $1 if a female soccer fan in England sees it. Advertisers don't need to know my name, just how closely I fit their target audience. That might be very broad – males in the UK aged eighteen to fifty-four – or extremely specific – goths from the Midlands who no longer live at home, have recently paid a large phone bill, and have a specified credit rating with Experian.

Facebook's service to advertisers offers a range of methods for identifying the target audience. If you already have a mailing list – perhaps you're a political party with a list of supporters – you can simply tell Facebook to target those exact people. If you want to target anyone who visited your website that day, you can ask Facebook to track those visitors, using a pixel that identifies the machine and web browser used to visit that site. Either of these methods delivers a Custom Audience.

If you have a small custom audience, and you just want more people like them, you can ask Facebook to run its own algorithm to identify what your audience has in common but doesn't share with most other people. In effect, an AI program will treat your existing audience like a natural cluster, and then use the distinguishing characteristics of that cluster to identify others who are similar. Those new people form what Facebook calls a Lookalike Audience. If you don't know the specific people, but you do know what they're like – their demographic categories and other relevant attributes – Facebook will let you choose from thousands of categories to describe the type of person you want. There are limits on what you can specify, and with good reason, as we'll see below.

The auction for Facebook's advertising space isn't just about who will pay the most. Facebook also takes into account the 'ad quality and relevance' to that user, and the 'estimated action rates'. In other words, Facebook does want the audience to feel that an ad is relevant and interesting, not just annoying, and they especially want the viewer to take action as a result of seeing the ad – most likely by clicking on it.[322]

Much of Google's digital advertising is cost-per-click (CPC) so advertisers only pay every time somebody bothers to click on the ad. This gives instant feedback on how effective the advert is in that time and place, to that person. That's the key to the political campaigns that ran different ad versions and adjusted them throughout the campaign. It also gives the ad auction algorithm, and the humans who program and adjust the algorithm, incentives to maximise the click-through rate. This has led to problems when recruiting for jobs in male-dominated industries. An ad shown to more men will get more clicks, so those ads get shown to more men, thus perpetuating the under-representation of women.

Researchers at Carnegie Mellon University and Berkeley created a program called AdFisher that sent thousands of simulated profiles online to browse, and studied which adverts they were shown. The profiles tagged 'female' were far less likely to be shown adverts for highly paid executive coaching work than the 'male' profiles.[323] The study was done in 2015, and since then, Google, Facebook and other online advertising companies have made a point of excluding sensitive categories like gender from the targeting options for jobs.[324]

In fact, companies are now using AI to attempt more inclusive job advertising. Not only do they carefully show adverts to the full range of the human populations eligible for the job, they're even trying to analyse what kind of language might subconsciously appeal to men more than women. Words such as 'lead' and 'analyse' are described as masculine, and 'support' is a feminine word, according to one study, which recommended using software that could code such 'gendered' words and suggest replacements.[325] I can't help feeling that this reflects a very stereotyped view of what men and women are like.

In spite of this, when researchers Muhammad Ali and colleagues at Northeastern University experimented in 2019 by running adverts for different jobs to the same target audience of undefined gender or ethnic background, they found that Facebook showed the adverts to very stereotypical segments of that audience. Lumberjack jobs were shown almost exclusively to men, mainly White men. Supermarket jobs, by contrast, were seen by four women for every man who saw them. Changing the image in the advert to show Black or White, or male or female, models, had little effect on who saw the ad.[326]

As the researchers point out, the program may be choosing an ad audience based on characteristics that another Facebook program has inferred, in much the way that Twitter inferred that I was a man. They also note that in 'traditional media, an individual interested in seeing ads targeted to a different demographic than they belong to has merely [to] watch programming or read a newspaper that they are not usually a target demographic for.' Anyone can read a tech magazine and see the same adverts as every other reader, but when online ads are chosen for you, you may not even know what you're not seeing. 'On Facebook,' as the researchers note, 'finding out what ads one may be missing out on due to gender, race, or other characteristic inferred or predicted by Facebook is more challenging.'[327]

Advertisers like personalised adverts for several reasons. Nineteenth-century American businessman John Wanamaker is credited with the saying, 'Half the money I spend on advertising is wasted; the trouble is, I don't know which half.'[328] For advertisers, the promise of digital advertising is that they can finally know which half of their advertising spend is wasted, and stop wasting it.

When I interviewed PepsiCo's Chief Insights and Analytics Officer Stephan Gans for *Significance* magazine, I asked him to estimate the impact of data analytics on his company. 'I think we spend $2 billion a year just on advertising,' he told me. 'Say that you're 10 per cent more effective in targeting the right consumers, and convincing people to buy your brand: you're talking about saving millions and millions.' The promise of digital advertising is that it can measure precisely which adverts are the most effective, to whom, and when and where.

Advertising businesses talk about attribution – working out which adverts turned somebody into a customer. But it's hard to attribute results when a decision may be the result of lots of factors. Perhaps somebody saw a television advert that aroused their curiosity, became generally aware of a promotion through billboard posters, but only took the step to buy when, bored on a train, scrolling social media, they saw a twenty-four-hour sale on the very thing they fancied buying.

Omni-channel marketing, a campaign co-ordinated across different media, raises a challenge for attribution, as most of us don't remember what we've seen. Being able to track each individual and cross-reference what adverts they saw on television or online, or even which billboards they passed, would enable direct comparison of different paths to assess what was advertising money well spent, and what could be cut with little effect on sales. Advertising companies that sell billboard space do, in fact, partner with mobile phone companies to get information not just about who passes by their sites, but what other ads they're looking at online as they pass.[329]

Advertisers hope that a personalised ad is reaching a receptive audience, and therefore that money spent on it is in the half that is not wasted. They also want attribution to be as accurate as possible to keep improving the return on their advertising spend. But advertisers are not just thinking of isolated sales. They talk about customer journeys and 'customer relationship management'. Linking social media accounts, devices and email addresses with one identifiable, real-world individual is vital to building that relationship. If your favourite fashion brand or stationery store offers you a membership scheme, it's not just a cynical attempt to capture your data for their e-mailing list. It *is* designed to collect that data, but it also wants to keep you actively engaged, so you keep coming back for more. A customer who plays an active role in shaping that relationship is more likely to feel, not just recognised, but empowered.

Facebook and Google (Meta and Alphabet, as we should call them now) dominate the digital advertising market. Both companies have a very large database of profiles and a panoply of ways to link them to the person for whose attention the advertisers are bidding at that moment. Millions of apps and websites with no apparent connection to Facebook are feeding data to Meta from the moment they are opened, without the user's knowledge or consent. Google's many services – maps, search, mail, the Android phone operating system – are ideal for collecting real-time data on what a user does, building a profile better than a thousand consumer questionnaires. Their automated tools make personalisation available on a smaller, cheaper scale than would have been possible before.

There are, however, doubts about how much difference behavioural targeting makes. Advertisers have begun to suspect that advertising only to the people closest to their current customer profile probably means . . . only advertising to existing customers. Brands whose value to purchasers is partly about how others will see them when they wear Adidas, drive a Ferrari or drink Fairtrade coffee need to advertise not only to those who will buy their goods and services, but also to the audience watching their customers. Much of the joy of driving a Ferrari, lovely cars though they are, is being seen to drive a Ferrari. In order for that social validation to hold its value, the brand needs to be seen as desirable by the vast majority of people who will never get closer to ownership than buying a Ferrari-branded hat. Yes, I do own a Ferrari-branded hat.

The idea that each of us gets a completely individual selection of adverts is exaggerated. Most advertising is very broadly targeted. Hyper-personalisation – microsegmentation – is too expensive for the very niche advertisers who would benefit most from it, and largely not worth it for large companies who target large populations. Advertisers like to believe they're getting a hand-picked audience. We like to believe we're getting hand-picked adverts. The reality is probably closer to the dogfood my stepmother's terriers Tipple and Pickle have in their named bags. It's not one-size-fits-all, but it feels more personal than it is.

Companies, seeing that both legal regulation and public disquiet are creating new obstacles to collecting personally identifiable data, are also developing other ways to identify the right audience for an advert.

One method is contextual advertising that simply matches an advert to the web page or app, the time and location, and where it will appear. For instance, imagine that it's lunchtime and a mobile phone is moving towards a fast-food takeaway. That's the perfect opportunity for that phone to display a special lunch offer for that takeaway.[330] It doesn't matter whose phone it is. Digital adverts for Pimm's, the British drink most consumed in – and strongly associated with – summer, were programmed to activate when the temperature reached twenty-one degrees Celsius, and potential customers, whoever they were, would be most receptive to the thought of a long, cool drink.[331]

Another method is to move away from centrally collecting large banks of personally identifiable data, and towards distributed software systems that can draw conclusions without moving data off the device that collected it. Apple make a big point of being privacy-protecting, and not taking data off the user's phone or computer. Google, whose Android software is the most widely used on mobile phones, is moving in the same direction. This doesn't mean that we will no longer be profiled and targeted with personalised adverts and services. It just means that extracting data for mass processing elsewhere is no longer necessary in order to sort us into segments and microsegments. Your phone, or other device, will do the work in-house.

Personalised advertising is the least personal thing you can imagine. It's entirely automated and based on data, which is automatically collected and compared to thousands or millions of other data profiles. But it's also the template for other forms of personalisation. If you search for flights, you may find that using a different machine, or even logging out of your airline's membership account, will show different prices for the same flight. That's because the automated online systems are predicting the price you would be willing to pay, based on measures like what machine you use to browse, how close to departure you are looking, and how many times you have looked for the same flight already.

That's differential pricing. Is it fair to offer people different prices for the same product? Airlines would argue that they would rather not fly with empty seats, and their policies are designed to predict what a flight is worth to a particular person.

The travel planning app I described in the Introduction gives me options based on my departure point, destination and arrival (or departure) time. I can adjust my preferences if I don't want to go by bus, or prefer fewer changes to faster journeys. This is more convenient. It makes using public transport easier and less stressful, which encourages me to regard that as my first option for within-city journeys. But it's not geared only to my convenience. It often shows cycling or walking options at the top, even for journeys that would take 169 minutes to walk (and burn 700 calories, the app helpfully points out). The app was not designed by me, and I didn't pay for it, so I get the app that is designed to get me to do what somebody else wants me to do.

'When you deal with human persuaders, you can stop the persuasion process and ask for clarification, you can argue, debate, and negotiate,' as B. J. Fogg puts it in his 2003 book, *Persuasive Technology*. 'By contrast, when you interact with computing technology, the technology ultimately controls how the interaction unfolds. You can choose either to continue or stop the interaction, but you can't go down a path the computer hasn't been programmed to accept.'[332] As mentioned in Chapter 4, Fogg invented the term 'captology' to describe the study of Computers As Persuasive Technology, though he now prefers to talk about 'behaviour design'. Many of Fogg's students at the Stanford Persuasive Technology Lab went on to found technology companies, including Instagram. In one class, he set 'make an app for Facebook' as a homework assignment, and some of those apps made their designers thousands of dollars.[333] Still teaching at Stanford University, he is very concerned about the ethics of using machines to influence human behaviour.

In some ways, Fogg's early work was very prescient about how much 'persuasive technology' would infiltrate our lives within a few years. At a time when the laptop was still a luxury item, he predicted that interactive technology would be present in 'cars, kitchen appliances, perhaps even clothing', and 'the smart products and environments of the future won't just be about productivity or entertainment; they will also be about influencing and motivating people'.[334]

In other ways, Fogg underestimated how readily we would embrace this ever-present guiding hand. Even though he identified mobile technology as an important medium for future persuasive technology, he dismissed some possible applications. 'The media and some futurists have proclaimed the coming day when advertisers will push context-relevant messages to our mobile phones: as we walk by the Gap, we'll be offered 10 per cent off on khaki slacks, or as we near the bakery we'll get a message that cinnamon rolls are fresh out of the oven. My lab members and I are skeptical of this vision,' he wrote in 2003. 'Will people accept such intrusions into their mobile devices? We suspect not, unless the person gives permission to receive such information from specific sources.'[335]

As we've seen, precisely this kind of targeting, combining location, time and profiling, is in use today. You could argue that we are not always given the choice to say no to receiving such intrusive messages, or that the choice to opt out is made difficult and confusing, and I would agree. Given the choice, we often *do* opt out of intrusive targeting, when we can navigate the sneaky design of the persuasive technology.

But Fogg himself identifies one reason why we keep picking up our mobile phones without even thinking about it. Even busy people, he says, have 'moments of downtime' when they are essentially waiting for things to happen to them: waiting for their train to reach its destination; waiting in a queue; waiting for a friend to meet them. They're all classic moments when the easiest thing to do is reach for a mobile phone, which I nearly always do, even though I always have a book in my bag. In Fogg's words, I 'can turn to mobile technology to fill the void'.[336] It gives me a sense of some control in situations where I have no control, he notes. And, I would add, it makes me feel connected to other people I know, even when I am surrounded by strangers, plugging an emotional void.

I could, of course, start up a conversation with a stranger (when I do this it's almost always fascinating and life-enhancing). But that risks social awkwardness, and takes an effort on my part. Scrolling social media, or the news, or funny videos, on my phone is much easier, and that is by design. The technology is designed to be easy, inviting, rewarding – but not so rewarding that I feel completely satisfied and put the phone back in my pocket for the rest of the day. The combination of technology with psychological research that draws heavily on behaviourism has shaped our everyday interactions, with each other as well as with companies and institutions. Advertisers, transport planners and political campaigners all use technology to persuade and influence us.

We may harness these psychological effects to try to change our own habits. If I schedule a gym session with a timed reminder, I am much more likely to get to the gym. For stronger motivation, I could sign up for a scheme that rewards me with discounts or cinema tickets for achieving a certain level of exercise. Stronger still, I could invest in a 'Pavlok', a wearable device for my wrist, which can be programmed to give me an electric shock if I bite my nails or smoke or spend too much time on social media.[337] As we learn the language and social rules of digital communication, we also try to influence and persuade each other, as humans always have. The difference is that we don't design the technology, or hire the designers, so we are always working within structures and rules we didn't make, and of which we may not even be conscious.

The point of these systems, as you've gathered, is not just to offer us services and commodities that we're more statistically likely to use and pay for. It's also to make us feel that what we're being offered has been selected for us because of our particular qualities. Being shown adverts for sophisticated restaurants, or requests to donate to environmental charities, is both a social signal and a reward. The reality, as you're already aware even if you haven't thought about it this way, is that the world is not being sorted for our benefit. We are being sorted for the benefit of nameless others.

It's good to go to a buffet where plates are clearly labelled – vegetarian, vegan, gluten-free – because it makes it easier to choose what suits you, as an individual diner. It can also make you feel that your individual needs are respected. No more asking apologetically whether there is anything without bacon, because your question has been anticipated and answered. The promise of the personalised century is that your buffet will be not only labelled, but automatically sorted. No more walking past plates of food you can't or don't want to eat. To paraphrase the Thread advert: 'Imagine a buffet that only serves food suited to you. Well, now you don't have to.'

The reality is that you are the buffet. When it comes to personalised services, you are served the way dinner is served. The labels on the dishes – vegan, kosher, gluten-free, doesn't like beetroot – are describing you, not the food. The algorithm that uses Real Time Bidding to decide which advert you see is not bidding on your behalf for the most relevant or pleasing advert that you would choose for yourself. It's the advertisers who are bidding, and your value to them goes up or down depending on how statistically likely you are to respond to an advert, and what it's worth if you do.

This explains why I used to get adverts for powder furnaces on Twitter. It wasn't very likely that I would buy a powder furnace, as I'm not sure what it is, but if I did, it would be a big, expensive purchase. And it's not that likely that *anyone* would buy one via Twitter. As somebody tweeting about engineering and motor vehicles, I was slightly more likely to buy a powder furnace than somebody picked completely at random. I'm not seeing the adverts that most closely fit what I want or need. I am seeing the adverts *for which I am the closest fit*, or starker still, adverts for which my *profile* is the closest fit – a profile that isn't me, that I didn't make, of which I'm largely unaware, that includes things that have been wrongly inferred from available data.

Imagine this movie scene: midnight in the secret research

facility. Stealthily, you move through the dark corridors, slip through a ventilation shaft into the basement, and land with catlike feet. The beam of your torch flickers across computer screens and shining metal, then stops as you see what you came to find. Your own face, your body, a perfect replica of you, breathing as you breathe, moving as you move, suspended in a vast grid, one node in a web of countless others. Mindless, soulless, made from billions of datapoints you shed in every waking second, and perhaps some sleeping ones too. Like the mythical youth Narcissus, you gaze at your reflection, but not with love. Your skin prickles in horror as you realise: this is the replacement for you, being prepared to take over all your social roles, its future predicted from past data and by comparing it to all the other profiles hanging in rows, infinitely receding into the darkness.

You may be rolling your eyes now, and thinking, 'Oh come on, Timandra! You think we didn't know how our data is used to offer us stuff that people want us to buy or do or think? We're adults, we know our way around this world. And, for all the annoying times it's creepy or prejudiced or just gets it wrong, we like the level of choice it brings to our lives.' If you're thinking that, you're in good company.

Trevor Phillips bestrides two categories of work that are central to this book. In 2003, he was made head of the UK's Commission for Racial Equality, and when that became the Equality and Human Rights Commission in 2006, he was its chair for six years. In 2018, he became chair of Index on Censorship, an organisation that defends free speech internationally. He also runs a data analytics company, Webber Phillips, with the same Professor Richard Webber who pioneered using postcodes to classify people socially.

In their London office, I interview Trevor about how data and working for a better society can go together.

Several people accost him with questions as he walks me through a large open-plan space to the kitchen, makes me a cup of tea, and invites me to pour milk to my own taste. The names that a company gives its meeting rooms say a lot about it, which is why they're carefully chosen. We pass Mandela (for Nelson), Parks (for Rosa), and go into Lamarr, named (I assume) for Hedy Lamarr, the Hollywood star whose engineering research also laid the foundations for Wi-Fi technology. But if the meeting room is in typical tech company style, Sir Trevor Phillips, OBE, is not. Wearing a crisp white shirt with tie and cufflinks, he occasionally jots a word down with a fountain pen as we talk, or sets the pen gently spinning as he considers an answer. Trevor is driven, not only by a scientist's curiosity about the world, but also by the political campaigner's desire to change it. That's why I want to hear from him the positive case for using data to sort, profile and target individuals.

'I think it's almost wholly a positive story,' he tells me. 'For the first time, as far as I can make out, in human history, it is possible to think seriously about how people's individual desires and preferences can be answered without having to suppress everybody else's desires and preferences.' Unlike the physical world, where we have to compromise our individual preferences to live together, the digital world can address each person's wishes as if they were the *only* person in the world.

'When it comes to this issue of equality and diversity, I think that's a good thing,' says Trevor. 'Because human beings are different, groups are different and individuals are different, and that's what makes life interesting. The more we're allowed to be ourselves, the more we can realise our own potential, I think the better it is for everybody. I think the idea that digital technologies provide more options, and therefore more choice, leads to the liberation of human potential.'

This is, indeed, the promise of self-realisation that the personalised world offers. Whatever is the best fit for you – and people very like you – is what you get. But doesn't that also mean accepting the world of division and discrimination as it is? Trevor doesn't agree.

'I think we've got ourselves into a bit of a mess here, where somehow we think that satisfying the precise needs of groups of people, because of what their preferences are – and they might come out of historical or biological or cultural reasons – will somehow eventually lead to people being discriminated against.' He uses the example of television. 'You know, when we had three channels, you had to compromise and get the middle of the road that everybody else had, and there'd be a few highlights everybody wanted to see anyway, Morecambe and Wise, or a cup final, or whatever.' That was before Channel 4 arrived with its mission of catering for neglected minority audiences.

'Now, we've got a hundred channels, you don't have to compromise,' says Trevor. 'Is that a worse situation than the one where you have three channels? For most people, of course it isn't. I think the point about all of these technologies that provide more options is actually a) recognition of diversity, b) better, greater satisfaction for the individual.'

Webber Phillips uses two important tools to profile groups of people. Alongside, and often combined with, the geodemographic methods that Richard Webber pioneered, their Origins program predicts ethnic origin from family name. This goes far beyond the broad categories of 'Black' or even 'Nigerian'. It can point to particular regions or tribes in the family's country of origin. The company does have commercial clients who use such information for marketing, but it has social purposes too. Applying this method first alerted Trevor Phillips, and thus the UK Government, to the disproportionate impact of the first wave of Covid-19 infections on particular communities. Trevor noticed that the hardest-hit neighbourhoods were the same ones Webber Phillips identified as having large numbers of residents belonging to ethnic minorities. Although the system works from individual names, Webber Phillips will not allow it to tag individuals as coming from a particular ethnic origin. Its use is purely to characterise groups by their ethnic mix.

That group doesn't have to be a geographic population. 'We did a nice project for the Barbican, which analysed half a million bookers over a period of time,' says Trevor, who is also a member of the Barbican arts centre's board of trustees. 'What we discovered was that people from Japanese backgrounds who go to classical music a lot tended to book one or two tickets. South Asians tend to book six or more. Because for the first group, it is Anne-Sophie Mutter playing the violin. And they want to see her play the violin. They don't care about anything else going on. That's what they're paying for. South Asians – it's an evening out. And it happens to be Anne-Sophie Mutter playing the violin because little Ali is learning the violin. And Auntieji is going because she wants to be part of the family group, and you can't go out without her.'

Trevor's example is a generalisation, he says, but it provides a social insight that's useful if the venue wants to attract more of particular groups. 'In the South Asian group, let's say, you've got to make it a nice evening out. Whereas for the Japanese group, they're going to come and go. And we could only do that by analysing patterns of booking.'

There is a big question that has been bubbling up in me as we talk and, if I'm honest, for as long as I've been thinking about what's in this book. Data is very good at capturing what's observable from the outside. But if the point of this new society of choice is its potential for self-determination, for each of us to live exactly the life we want, and be the individual we feel ourselves to be, isn't it wrong to label us from the outside, without even asking us who we think we are?

Trevor sets his fountain pen spinning and then answers thoughtfully. 'I think the first question you've got to ask yourself is why are we asking the question about categories? And it matters in relation to sex, gender, race and religion. Because what you think you are may not be the thing that matters.' A bold claim, which he follows with a specific example. 'We did an analysis of about a million patient episodes for a local health authority. We got the technology that uses the names to analyse a sample, by ethnicity, language, origins, etc.' Meanwhile, the same patients had been asked to identify their own ethnic category. And the answers did not line up.

'Sikhs are a good example,' says Trevor, 'because it's very rare for somebody who's not a Sikh to have a Sikh name, because the name is part of the identity. About 30 per cent of the people with Sikh names had ticked a box which identified them as White British. Why? Well, when they tick the list, we think they were looking at the term British, which is what they want people to think of them as being. And they probably never got down to the bit that says,

"Are you Asian Sikh?"' So the people Webber Phillips categorised as Asian Sikh identified themselves as White British.

'Now, that's all useful information,' Trevor continues. 'It tells you something about how people want to be addressed, how they want to be thought of. Where it's not useful is when you're dealing with data on pay gaps, for example, because the point of establishing pay gap information is to work out whether people are being denied opportunities.' The purpose of categorising people in this case is to look for wider patterns of inequality, work out what causes them, and try to change them.

'By and large, nobody gets denied an opportunity because they think in their head, they're Black or Sikh,' Trevor asserts. 'They get denied an opportunity, because everybody else thinks that about them. We're asking people, and we're categorising people, according to what they think they are. But actually, what matters is what everybody else thinks they are.'

This means, of course, that how we categorise people may be different for different purposes. I'm reminded of a discussion I chaired for the Royal Statistical Society's Data Ethics group, about collecting data on sex and gender. It revealed to me how important it is to be clear what question you are asking, and for what purpose. Information on biological sex and on self-identified gender are useful for very different reasons, and a lack of clarity about what is being asked, and why, can render the data useless for either purpose, as the Office for National Statistics discovered after the 2021 Census.

'We have a set of characteristics, each of which we share with some other people,' as Trevor puts it. 'The sets of people may over-lap a bit, but they're not always the same. What makes us really interesting as individuals is the peculiar combination of those characteristics. For me, my colour probably is a big negative. But my educational background is a massive positive.'

Trevor studied chemistry at Imperial College London, the science equivalent to Oxford or Cambridge, where he got his taste for testing ideas against data. 'I think this idea that each of us is a peculiar, unique configuration made up of these different characteristics is really important. Sometimes one bit of a configuration confers disadvantage. And that needs to be dealt with in some way. But you can't really understand when it does confer disadvantage, on average, unless you've got that data.'

Trevor gives me a personal example. 'I can make a great speech about how I would be in a different place if I were not Black. But the truth of the matter is, I'm incredibly fortunate. On the whole, there is no basis for me to go around saying I'm in terrible state because of my race. But for six months or so, when Black Lives Matter happened, irrespective of what I think about myself, that became easily the most important thing about me. And I think for every Black person, suddenly, that was the thing that mattered about them.'

Whatever Trevor thought about himself, the fact that others immediately saw him as Black, in a context where the relations between Black people and the rest of society were a live social and political issue, would affect their interactions. This brings me to my major worry about the computer systems that profile us. They also decide which categories we belong to, based not on who *we* think we are, but on data collected about us, our external characteristics. If the personalising algorithm, whether it's Netflix's recommender system or Google's advertising network, is choosing what it thinks I would like, based on a cluster or category that it's allocated me to, isn't that narrowing, not widening, my choices? Trevor Phillips worries about that too.

'When I was working in the United States a lot about four or five years ago, I started getting these ads for bail bondsmen. Google had looked at the things that I was interested in, and it clocked "African American male of a certain age". So, I was getting these ads.' The algorithms had inferred characteristics that limited the options that Sir Trevor Phillips, OBE, would see online. This curating of options to our inferred characteristics risks nudging us towards becoming more like the cluster to which we were allocated, thus fulfilling the algorithm's predictions.

'You're right, this is a very big danger,' Trevor agrees. 'It carries the additional danger, as you say, rather than expanding our choice, [of] narrowing our choice. But I think when the technology works, that's a good thing. Because it means I'm not overwhelmed by choices I don't want, and I'm never going to want. Spotify probably doesn't show me things that I *would* like. But it shows me things that I *do* like, and, frankly, most of time, I'm happy with that.'

When the technology narrows down the options we see, Trevor suggests, it's often because of our own behaviour – we don't click on an ad, or we stop watching the video before it ends – or because it doesn't have enough information to profile us accurately. 'I'm actually of the point of view that I want the system to have more information, rather than less, to know me better, as opposed to somebody like me,' he says. 'But let's remember, we do have agency. Sometimes these arguments develop in a way that seems to forget that we can make a choice.

'So I'm quite relaxed, to be honest. I think we are, if we choose to be, led. I get that they may develop in a way that anticipates what we want. But these things are not static relationships, they're dynamic. And the machines are learning from us as much as we're learning from them.'

It's always tempting, as a writer, to pick a line and pursue it to extremes. I could have stopped when we discovered that personalisation through data is the least personal thing you can imagine, and that scene which I hope will appear in the Hollywood adaptation of this book, of all the digital doppelgangers swaying in the underground laboratory. That would have left me free to denounce the technology, and all who build it, as evil manipulators, and call on you, dear reader, to bin your smartphone, ditch your social media accounts, and run out into the forest in your bare feet to rediscover your true, authentic self.

But this isn't eighteenth-century France, and I'm not Rousseau.

On the other hand, I'm not as relaxed as Trevor Phillips. I don't entirely share his faith that personalising technology will open up a utopia of personal freedom and self-realisation, equal opportunities hand in hand with tailored services to meet particular needs. I do agree, though, with his point that we have agency.

The problem – and there is a problem – is not the technology. It's what technology people call a 'wetware problem'. It's not the software, or the hardware. It's us.

Chapter 9: It's Not the Technology – it's Us

'For modern people, the narrative form entails seeing one's life as having a certain arc, as making sense through a life story that expresses who one is through one's own project of self-making.'[338]

KWAME ANTHONY APPIAH

Personalising and profiling technologies work because very clever tech companies psychologically manipulate us, using our human frailties for their own ends. The most successful data-gathering, profiling and targeting technologies are the ones that make us willing participants, like the animal trapped because it won't let go of the biscuit in the jar.

They feed our desires for affirmation, recognition and human contact just enough to keep us coming back for more, whether that's social media likes, or search results, adverts and offers that seem to 'get' us and our unique desires. Thanks to the inbuilt feedback loops – offering us content, or activity, and then recording our responses as data – the programs are calibrated to keep the process going, endlessly refining our data profiles and experimenting with ways to predict and guide our actions. In that sense, we live inside a giant laboratory, eager little rats pulling the levers of our smartphones to get our digital rewards. No wonder critics call it addictive and compare social media use to slot machine gambling. We're not rats, though, and we're not in Skinner boxes. The wider contexts of our lives are important.

Much of modern psychology is based on getting real humans to participate in experiments in laboratories. That way, the researchers can apply scientific principles like comparing groups of people who get different treatments, to try to find out the effects of whatever it is – making them think of old people before measuring how fast they walk, for example, or telling them that they will be shown adverts based on their own browsing behaviour. The participants in these experiments tend to be psychology students, because those are the people easiest to recruit for small payments or course credits. In fact, only 20–30 per cent of experimental subjects in published psychology studies are *not* undergraduates in psychology courses. If all humans responded the same way to the same stimulus, this wouldn't matter, but we don't.

As three psychologists pointed out in 2010,[339] most published social psychology is based on experimental subjects who are WEIRD: Western, educated, industrialised, rich, democratic. Two thirds of them are in the United States. In an article called 'The Weirdest People in the World?'[340] Joseph Henrich, Steven J. Heine and Ara Norenzayan showed how poorly this homogenous group of college students represents supposedly universal human behaviour.

For example, they compared how easily people from different cultures were fooled by a common optical illusion, the Müller-Lyer illusion. You've probably seen it before: two lines of equal length, one above the other, with short lines forming arrows at their tips. One pair of arrows points outwards, the other points inwards. Even when you know that the lines are equal in length, it's very hard to stop your mind perceiving the pointing-outward-arrow line as shorter.

If you're WEIRD, that is. American undergraduates only saw

the lines as equal when one was about a fifth longer than the other. But the San foragers of the Kalahari in Africa saw the two lines, correctly, as equal. Other researchers have failed to replicate this particular finding, but many other psychology experiments get robustly different results in different cultures.

In the field of behavioural economics, too, supposedly universal findings turned out to be very different in different societies. A commonly used experiment called the Ultimatum Game gives a small sum of money to a pair of participants to share between them. Participant A gets to decide how much to offer participant B. The catch? If Participant B rejects the offer, neither of them receive anything at all. Frequently, A will offer 40–50 per cent to B, keeping 50–60 per cent for themselves, and typically, B will reject offers below 30 per cent, even though that means they get nothing, apart from the schadenfreude of knowing that stingy A gets nothing either.

Again, though, these results are only typical among WEIRD undergraduates. When researchers took the same experiment to different cultures, they got very different responses. Societies based on living together in small groups, subsistence farming or hunter-gathering, offered and accepted much smaller shares. Analysing the results, the researchers found that knowing the size of social groups, their familiarity with market exchanges, and whether they belonged to a world religion, all predicted how much participants would offer and accept. Later studies found societies where adult participants might even reject offers above 60 per cent for being *too generous*. This phenomenon, observe the researchers drily, 'is not observed in typical undergraduate subjects (who essentially never reject offers greater than half)'.[341]

Spatial awareness, navigation and attitudes to risk also varied between cultures, though some similarities seem to hold across both industrialised and small-scale societies, including social and emotional perception and expression, and the finding that men value physical attractiveness in a mate more than women do. But one psychological dimension does vary a lot between societies and cultures: the sense of self.

Compared with people in other societies, Westerners are more likely to think of themselves as independent beings, driven by internal, psychological traits, making autonomous judgements and acting as individuals. This self-view is linked with a tendency to over-estimate one's own personal qualities relative to everyone else, the desire to stand out from everyone else instead of fitting in, and placing a high value on personal choice.[342] Sound familiar?

People from non-Western cultures don't always respond in the same way to psychological experiments, and their behaviour points to very different ideas of who they are, and how they fit into their world. One reason for refusing overly generous offers in the Ultimatum Game is that, in the participants' shared culture, receiving such a gift puts the recipient under an obligation to the giver. Compared to the WEIRD population, people from East Asia and Malaysia, for example, are more likely to see themselves as part of a web of social relationships, a network that comes with obligations as well as benefits.

The vast majority of Western drivers think that their driving is better than most other people on the roads. In one study, 93 per cent of American drivers, and 69 per cent of Swedish drivers, said they were better-than-average drivers.[343] Ask American college professors if they're better or worse than their colleagues, and you get results that are just as impossible. One study even found that participants viewed Rasputin, the unsympathetic Russian monk, less harshly when they were made aware that they shared his birthday.[344] But this self-enhancing bias is not as strong in other cultures. The most individualistic group of people ever studied are American college students, and there is evidence that their trends towards individualism are getting stronger with every succeeding generation.

Psychology experiments published in peer-reviewed scientific journals are very likely to show what happens when you get a small group of the most individualistic population in the world into a laboratory, and study their behaviour under entirely artificial conditions. These are the assumptions that underlie the technology – largely designed by college-educated Americans – that permeates our lives. That's why we're wrong to think of the technology as targeting universal human needs and desires. It was designed to target needs and desires that are specific to here, to now, to us. Our view of the world, through the prism of our own identities, is what shapes the technology we have. But it's not a view of the world that's universal to all humans in time and space. It arose from our culture's specific history, as I described in earlier chapters.

And, despite the high levels of self-belief in WEIRD drivers, professors and people who share Rasputin's birthday, ours is not a society of confident, self-sufficient individuals. We are insecure, anxious, constantly thinking about ourselves, and how others see us. As Christopher Lasch observed, we're not rugged cowboys facing outwards to unconquered frontiers, but fragile narcissists who need to see ourselves reflected in the world.

A survey in 2019 found that almost one in five Americans

received treatment for mental illness in the past year, defined as medication or talking therapy from a mental health professional. The rate among women was one in four.[345] In the UK, the NHS reported in 2020 that one in six children aged between five and sixteen had a probable mental disorder. That rate rose to over a quarter in girls and young women aged seventeen to twenty-two.[346]

As social media use among young people has risen alongside rising rates of reported mental distress, many campaigners have pointed to social media as the cause of mental disorders, especially among young people. Psychology professor and author Jean Twenge links the increase in loneliness among teens with the rise of smartphone use in the same age group. Since 2010, she reported, fewer American teens meet up with their friends every day, and more report feeling lonely.[347] That study was pre-pandemic, so it doesn't even take into account the effects of lockdowns and social distancing.

However, though the era of the smartphone coincides with the era of increasing loneliness and mental distress, there is scant evidence that more use of social media is the *cause* of less in-person social time. Several researchers have found the opposite – that teenagers who use social media more are also spending more offline time with friends.[348] Young people do spend less unsupervised time outside the home than their parents' and grandparents' generations did. British children start to play outdoors without adult supervision aged eleven, on average – two years older than their parents did.[349] Increased online interaction may be partly compensating for lack of in-person socialising.

Psychology professor Peter Gray, founder of non-profit group Let Grow, suggests that the loss of independent play – children and teenagers exploring and socialising without adult supervision – is responsible for the decades-long increase in anxiety and depression among young people. He notes that 'in some respects – such as freedom to choose what they wear or eat – children have gained autonomy over the decades. What has declined specifically is children's freedom to engage in activities that involve some degree of risk and personal responsibility away from adults.'[350]

Dr Amy Orben, whose laboratory at the University of Cambridge specialises in studying digital media and mental health, has found that the effects of social media use vary significantly between girls and boys, and at different ages.[351] She has found that adolescents who used social media more also reported lower levels of life satisfaction, but points out that this doesn't necessarily mean that social media use decreases life satisfaction – it could be that teenagers who are less happy with their own lives spend more time on social media.[352] One study found that looking at social media in the morning was associated with lower mood later in the day – but only if the person was alone while checking in online.[353]

None of this means that, if you are a parent, you should allow children of all ages untrammelled access to everything online. Just as offline teenage life can involve bullying and unhealthy peer pressure, online life can bring the worst, as well as the best, of adolescent life into every waking hour, wherever a young person picks up a device. But it does suggest that getting rid of social media would not resolve the problems faced by either teenagers or adults. As Professor Chris Bail says, 'I don't believe that quitting social media is like quitting smoking . . . Instead, deleting our accounts would require a fundamental reorganisation of social life.'[354]

Chris Bail and his Polarization Lab wanted to know why social media fed the perception of polarisation, and social hostility towards the 'other side'. Like many researchers he thought that the answer might be to get people out of an online 'echo chamber' and expose them to a wider range of views. Instead, his experiments found that more exposure to strong opposing views often hardened political positions, because 'stepping outside the echo chamber was not creating a better competition of ideas, but a vicious competition of identities'.[355] Challenging somebody's ideas on social media, as in the physical world, is too often felt as a challenge, not to what they're arguing, but to *who they are*. 'We are addicted to social media,' Bail wrote, 'because it helps us do something we humans are hard-wired to do: present different versions of ourselves, observe what other people think of them, and revise our identities accordingly.'[356]

Forming a stable identity in our social context is a process of experimentation, especially in adolescence as we build independence from our family and find our own social groups. A vital part of identity formation is testing different versions of ourselves on the people around us. There's often an intense burst of this during the late teenage years and when young people leave home, perhaps to go to university. Haircuts and colours, clothes, names, self-expressed sexual orientation and gender identity often go through rapid changes during this time. Social media is one location for this. How others respond to these different versions of the same person is a vital part of shaping a stable identity. Learning that others don't see you exactly as you see yourself can be unwelcome, but it's part of maturing. But another part of shaping a stable identity is to find a place in society where you feel you belong, among others who accept you as one of their own.

Imagine for a moment that you are St Alexis, that early Roman saint whose story we found on a fresco. You come home from a long journey, and nobody recognises you – not your parents, not your spouse, nobody. You go to the places where you used to hang out with your friends, and they ignore you. In the whole city of Rome, which you used to call home, not one person knows who you are. I'd spend the whole day saying, 'It's me – Alexis!' and having joyful reunions. I wouldn't be able to stand the loneliness of being invisible to people who used to care about me. I would start to feel that I didn't really exist. The saint's embrace of that anonymity, by seeing out his days as an unknown servant in his own home seems to me like an extreme form of self-punishment, as agonising as the physical torture that other early saints willingly suffered.

Feeling that you are part of a group, that you have a *shared* identity, meets our need to belong. The fear of losing that sense of belonging drives the extreme reactions that Chris Bail and his colleagues saw. When the volunteer subjects were exposed to strongly opposing views, they felt it as an attack on the group with which they identified. Their loyalty to the group hardened. Their willingness to listen to other views closed down, and their perceptions of the 'other side' became more negative. This pattern, regarding politics as a matter of identity rather than an exchange of ideas, did not originate on social media, as we saw in Chapter 3. But it flourishes there, precisely because the world of social media is so perfectly geared to rehearsing and performing identity.

Imagine a person living in a non-WEIRD society, perhaps in a small subsistence community today, or a few centuries ago in pre-industrial Europe. Imagine asking them, 'Who are you?' Their answer would place them in the fabric of their human world – their family ties, their work, perhaps their status in a relatively stable social hierarchy. They have answers ready to hand. Modern surnames often derive from those reference points – Johnson was the son of John, Fletcher belonged to the family who made arrows, Harkness meant you lived in the Scottish village with the chapel on the headland, so I'm told.

Now ask yourself the same question: who are you? Do those answers about family ties, work, community, feel sufficient? Perhaps it depends on who is asking you. You're unlikely to live in a world where those connections knit you into one stable network. That's why you need to express who you are to the world, to create and recreate your online persona, to find out where you belong. Identity is an issue today because how each of us fits into our world has become an open question, one to which we have to seek answers as individuals, free and anchorless in this world of boundless choice. Because we can choose from so many possibilities, we're always aware, on some level, that we could have chosen to be somebody different.

That's why we keep seeking affirmation of who we are. As social creatures, our sense of self is always in relation to others. As WEIRD, modern people, our sense of self is also supposed to be an expression of our true identity. It must be authentic, innate, an inner kernel, and also expressed, created in outward form by ourselves, showing others who we are.

This outward expression demands a response from others, a recognition of the self we're trying to express. 'I hope you can see me like I see myself,' as singer Sam Smith put it so well in that 2019 Instagram post. This, not the technology, is the root of the problem. If all the apps and smartphones and data brokers in the world were made illegal tomorrow, our lives would get very inconvenient, and that root problem would remain. In a world of individuals, unmoored from the social ties that confined us but also gave us a place in the world, we will continue to seek affirmation of who we are until we find other ways to feel secure in our own skins.

The endless search for recognition is not, after all, confined to the digital world. Politics, fashion, everyday interactions at work and even in private are all suffused with questions of identity. Blaming technology, and the companies and people who create the technology, is not just a misdiagnosis of the problem we have, it's a misdiagnosis that makes the problem worse.

Until the late nineteenth century, doctors prescribed bloodletting for a wide range of conditions. US President George Washington, after riding in the snow, developed a fever and breathing problems. He was treated with bloodletting and laxatives and died from a mixture of throat infection and shock. What was a standard treatment for thousands of years was just making patients weaker.[357] Regarding ourselves as the helpless victims of evil tech companies does a similar thing to our sense of self.

A wise friend once listened to me explaining why I couldn't do something – I was hopeless at sport, or at getting up early, or whatever – and replied, 'Argue for your limitations, and sure enough, they're yours.' Much later I discovered she was quoting author Richard Bach, though the saying has become a kind of proverb because it captures the easy slide from self-awareness to self-fulfilling prophecy. When I learned to ride a motorcycle, my lack of natural aptitude meant I took nearly seven days to do a five-day intensive course. Early on in the process, the very encouraging instructor said, 'There's no such thing as a natural, and no such thing as a person who can't do it.' Patiently, he overcame my difficulty co-ordinating all my hands and feet while balancing a two-wheeled machine. When I passed the test on my second attempt, he shared my sense of achievement. 'Well done,' he said, 'because you're not a natural, are you?'

If we think of ourselves as people who are at the mercy of our own mobile phones, who can't resist the evil wiles of the Silicon Valley tech companies, that's who we are. If we think of ourselves as persons, with agency, with the capacity to determine our own futures, that's who we are. The identities we choose for ourselves do, in part, shape what we do. The problem is that our current self-understanding *through* identity, no less than a persona built up from data by machines, reduces what we are to a matrix of characteristics, predictably reacting to what is done to us.

Philosopher Kwame Anthony Appiah has thought a lot about identity. He says the community in which he grew up 'was Asante, was Ghana, was Africa, but it was also (in no particular order) England, the Methodist Church, the Third World: and, in his final words of love and guidance, my father insisted that it was also all humanity'.[358] He now lives with his husband in New York.

Appiah agrees with John Stuart Mill, that liberty is important because each person needs to develop individuality. It's a good thing if each person can make their own life, regardless of whether that life is what others would call a good life. But our modern ideas of identity throw up some contradictions, which Appiah describes as two ways of understanding individuality. One is identity as an expression of the authentic self: the ability to live in the world as the person you truly are; to discover the *real* you inside, and insist that the world respect and recognise it when you express it outwardly. That's the way Rousseau and the Romantics would understand identity. The other is to see your self as a project that you create from scratch, completely free to choose who you are (but not free to *not* choose). Appiah doesn't think this abstract individual is either possible or desirable, though it's an idea that sits at the heart of what he calls liberalism – a political and philosophical tradition built around free and equal persons with moral autonomy.

'An identity is always articulated through concepts (and practices) made available to you by religion, society, school and state, mediated by family, peers, friends,'[359] he says. Or, to put it more directly, 'Autonomy, we know, is conventionally described as an ideal of self-authorship. But the metaphor should remind us that we write in a language we did not ourselves make . . . we may shape our selves, but others shape our shaping.'[360] This, I think, is both a timeless description of the human condition and a crucial observation on the ambivalent role that identity plays today, as a way of understanding our relationship with the world, with each other, and even with ourselves.

'To create a life is to create a life out of the materials that history has given you,'[361] says Appiah. Growing up with Ghanaian, English, Asante, Methodist roots; being same-sex attracted; having a political outlook partly shaped by his father's activism for Ghanaian independence from the British Empire: all these were materials from which he has made his life as a philosopher and writer in America. Because of this, he also rejects the first description of identity as an authentic inner kernel that just needs others to get out of its light in order to grow into itself. We are given the material, the marble, from which to carve ourselves, but the form in which we emerge depends on what we do and how we live.

Freud described the human mind as made up of different

elements that are often in conflict. The Ego, or executive agency, is partly conscious and partly unconscious; the Superego influences and censors the Ego, repressing instinctive drives and representing internalised moral values; the Id is wholly unconscious, driven by the pleasure principle, sexual and aggressive urges. I like to think of the Ego driving the car with a mixture of subconscious (changing gear and using the mirror and indicators) and conscious (reading the road signs and remembering the route) processes, while the Id screams from the back seat for ice cream and a toilet stop, and the Superego reminds the Ego to stay within speed limits.

Whatever you think of Freud, this divided self rings broadly true to me. I have internal dialogues in which I berate myself for being so lazy and disorganised, and ask my subconscious what the hell it was playing at when it put into my mouth the sharp reply that reduced my friend to tears. My subconscious is much smarter than I am but has no morals at all.

Certainly, the human mind is not always unanimous – 'of one soul'. We want many incompatible things, for reasons that can't easily be weighed side by side. We sometimes hate ourselves for having thoughts and feelings but act on them anyway. Sometimes we are only aware of making a decision when we act. When I make up my mind, I am not only choosing what to do, but which person to be. Am I the person who takes the job working for a charity abroad, or the person who takes the job living near her ageing parents to care for them better? We'll only find out when I take one job or the other (spoiler – I am neither of these people).

When Appiah talks about making a life as a two-way process with making an identity, he means something like this. Our beliefs, values and loyalties, as well as our circumstances, shape the life we choose to live, and the life we choose to live shapes the person we become. This has always been true, but for most people in most of history, the choices open to them about what life to live were very limited. The degree of choice open to most of us in the developed world today makes the question of what life to live – not merely how best to live the life that fate has allotted us – a real question. What is a uniquely modern, and WEIRD, dilemma is the degree to which identity is both the form in which we author ourselves, and a constraint on that self-authorship.

Writers often talk about having two heads, the writing head and the editing head. The writing head gets words down, the editing head sees what's wrong with those words and suggests changes. The writing head sulks a bit, and then rewrites, and both agree the outcome is better. Obviously, when a writer works with an actual editor, that relationship happens between two humans.

Ernest Hemingway never did advise writers to 'Write drunk, edit sober.' It's a misattribution of a line from a novel by Peter De Vries, *Reuben, Reuben*. 'Sometimes I write drunk and revise sober, and sometimes I write sober and revise drunk,' says McGland, a character possibly based on poet Dylan Thomas. 'But you have to have both elements in creation — the Apollonian and the Dionysian, or spontaneity and restraint, emotion and discipline.'[362] Too much spontaneity and emotion produces an unreadable, self-indulgent mess that makes no sense to anybody else. Too much restraint and discipline stops any experimentation, no risks are taken, and nothing new can be created.

Being the author of one's own life, one's own self, also involves that internal process. I act, then I reflect on what I did, and perhaps decide it was wrong, and resolve to act better next time. But I try not to become so self-conscious, so paralysed by fear of making a mistake, that I become unable to act. How does my internal life editor make these judgements? I have internalised a web of moral and social norms, by which I judge others, and then I turn back to look at myself as if through another's eyes. But I don't only judge myself against what I might call universal standards. As well as my own conscience, I judge myself against the person I want to be, and the person I want to be seen as by others.

For example, a few years ago I made a radio series called *How to Disagree – A Beginner's Guide to Having Better Arguments* (you can still listen to it on Audible and Apple podcasts). It took as its premise that it's good to disagree, and that arguing is fundamental to human reason, let alone living together in diverse societies, but that we tend to do it badly. We take differences of opinion too personally, because we're experiencing them as an attack on our identity instead of an invitation to reconsider. We don't listen enough to what people are actually saying. We assume that anyone who disagrees with us must be stupid or misinformed or a bad person.

After the radio series went out, I became much more thoughtful about my use of Twitter. Like many people, I am apt to tweet in response to things that provoke strong feelings in me: joy, amusement, exasperation or anger. Having presented a radio series telling people we all need to argue better, to refrain from personal attacks or assumptions of bad faith, to listen and take others' points seriously, and respond in a way that invites an equally thoughtful reply, and to be genuinely open to changing our own minds, I was forced to admit that many of my impulsive tweets did not meet those standards at all. In order to be a consistent person, at least in public, I have become much more thoughtful on Twitter. I try to live up to the standards I professed. You could say that, having identified as a person who values high standards in public argument, that identity itself feels like a standard that I need to live up to, hence a constraint – a good constraint, let's be clear. If I didn't genuinely think it was important to have better arguments in public, I wouldn't feel constrained by the identity I had created for myself. I also wouldn't have made the series. It was a good example of creating the person you want to become by what you choose to do.

Trying to be a consistent person is, in many ways, vital to being a functioning human. We all have faults, but if our faults are a predictable part of our character, we can make allowances for each other. My friends learned over the years to make allowances for my pathological lateness. They would tell me a meeting time half an hour before everyone else, knowing that I still wouldn't be the first to arrive. It was annoying for them, but they accepted it as part of who I am.

When, as part of another radio programme, I was asked if I had a habit I would like to change, 'being late' was the easy answer. I hated being late, I knew it was rude and inconvenient, and it was stressful for me too. I didn't set out to be late, but somehow, I always ended up on the last possible train that could get me there on time, but only if everything else fell into place perfectly. Which it didn't, obviously. I interviewed a couple of my friends about how I could change from being an always-late person to an always-on-time person. They didn't think that was possible. It was as much part of who I am as the colour of my eyes. I didn't think it was possible either. I interviewed a psychologist who specialised in changing habits. She explained that it's easier to make a new habit than to break an old one, with examples from dieting and smoking. Finally,

I asked her if she'd ever successfully changed a habit, and she told me that she, a lifelong late person, was doing all the things she'd told me to do.

It was like switching a light on inside my mind. If this psychologist believed that she could change, it must be possible for me to do it too, at least for the week I had to make my radio piece. I did all the things she suggested – plan my journeys the day before, instead of when I should have been leaving; build a fifteen-minute buffer into my journey plans; and, most important of all, decide that it was a priority for me to arrive on time. I would have to leave cups of tea half drunk and break off phone conversations, but I could do it if I wanted it. For the whole week, I was on time for everything. I left cups of tea unfinished at home, but sometimes I got a cup of tea at my destination because I was fifteen minutes early.

You're probably thinking I'm a bit pathetic, making such a deal about something so basic, but by the end of the week I was quite exhilarated. Not only had I arrived on time for everything, work and social, I had changed from being a late person to a punctual person. If I had managed it for a week, after all, there was no reason I couldn't continue to manage it. I have lapsed sometimes since then, at periods when my life was more generally getting out of control, but I no longer think of myself as a tardy person.

My friends found this change astonishing. They had to adjust their expectations and stop giving me an earlier arrival time. The new me was inconsistent with the old me, and that took some getting used to, after years of knowing me as the one who's always late. I should start some kind of therapy based on making radio programmes as the key to changing your life. It's certainly worked for me.

But it's not always so easy to get your family and friends to accept the new you. I have a friend who's a psychiatrist. She has a particular interest in something called functional neurological disorders. These are disorders which can't be explained by any medical cause, and are therefore treated as having a psychological origin. What Freud described as his patient Anna's hysteria – partial paralysis, extreme lethargy and inability to speak – would now be counted as a functional neurological disorder. This does not mean, as my friend makes a point of saying to me and to her patients, that the symptoms are not real. Disorders can include the inability to walk, speak or see, and be very disabling.

To successfully treat these disorders, my friend tells me, means getting inside the patient's mind to understand not only what symptoms they are experiencing, but what meaning they attach to them. It's always wonderful to me to hear how she has restored people to being able to walk, see, or whatever their disorder had taken away, by working with them through their mind alone. Somebody arrives in hospital unable to do something central to their life, and she sends them home with that ability restored.

But that's not always the end of the story. After a few months or a year, the same patient may return with similar symptoms, still with no detectable physical cause. A home or family dynamic that had adapted around a person with some impairment in their ability to function was unable to adapt to that person with their ability restored. Instead, the patient adapted by finding the symptom coming back.

That's an extreme example, of course. Mostly, when we change, families and relationships adapt. Or they don't, and relationships break up. One partner entering a new field of opportunity and challenge, higher education perhaps, may precipitate a divorce when the other partner can't deal with them developing new interests, friends, capabilities. Parents may struggle to accept the adult their child is growing into. Conversely, people change and decide that the old relationship, or friendship group, no longer fits the new person they are becoming. For children, breaking away from their parents and finding more importance in their friendship group is a normal and vital part of growing up. This usually comes along with some rejection of their parents' beliefs and habits, though maturity often brings an appreciation that their parents are not as wrong about everything as they used to be.

The reason I'm reminding you of what you already knew – that people change, sometimes on purpose, and that the people close to them don't always like it – is this: an identity is something that stays the same.

In mathematics, this is literally true. Something called an 'identity element' is the thing that, combined with any other thing, leaves it unchanged. When you're adding numbers together, the identity element is zero. You can add as many zeros as you like to three, it will still be three. If you're multiplying, the identity element is one, for the same reason – keep on multiplying that three by one until everyone's got bored and gone home, it'll still be three. For people, it's more complicated.

We can't literally stay unchanged, identical to the self we were even yesterday, never mind as children. Our bodies change, age, repair themselves, grow muscles or moustaches, shed pounds or hair, or vice versa. Our minds – our personalities, habits, beliefs, skills – also change as we interact with other people, and with the world, as we do things and things happen to us.

Remember Michel de Montaigne, with his fistful of water? He thought it a futile act to try to fix what a person was, in the ever-changing flow of life. All we could do, he thought, was try to express what, at any moment, we were thinking or experiencing. He would perhaps have had some sympathy with those who talk today about fluidity of identity, refusing to tie their identity to any constant. But an identity, by its very nature, must be expressed, in order to be recognised by others, and the act of expression pins it down. However you express that identity, whether explicitly – 'I identify as . . . ' 'Speaking as a . . . ' 'My pronouns are . . . ' – or by the way you look and dress, or by voicing the values and opinions that show you're part of a tribe, you are laying claim to a part of your being that will, you declare, still be important tomorrow.

Joy dresses as a goth, with a goth hairstyle and goth makeup. She hangs out with other goths doing . . . whatever it is goths like to do. And she's adopted the name Raven because she doesn't think Joy is a good name for a goth. That's an identity she's chosen. I'm assuming that she didn't grow up within a tight-knit Midlands goth community that raises its children to be goths like their parents and grandparents before them. If they did, why would they have called her Joy? But now she has chosen that goth identity, it brings certain expectations from others. To go back to the self-authorship analogy, she has chosen a genre. When we see a new book by a thriller writer we like, we buy it expecting a thriller. Even if we like romantic novels, that's not what we're expecting from this author. There's a reason why authors who have made a name in one genre often adopt a pen name for work in another genre. We all know that goths are people too, and that Joy doesn't live her life 'as a goth' twenty-four hours of every day, but her appearance, and her hanging out with other goths, tell us something about how she wants us to respond to her.

What about the other kinds of identities, like the ones Kwame Anthony Appiah gives as part of his heritage – Asante, Ghanaian, African, English, Methodist . . . ? He didn't choose any of them. They are what he lists as the material given to him, from which to make his own identity.

Here's where I think the way we understand identity, what it is, and how it expresses who we are, is crucial. For Appiah, his identity is what makes him *him*, and nobody else. This individuality, this ability to author his own life, using the materials life has given him, is completely central to his living a full, human life. I agree with him, both that autonomy is important to being a fully developed human being, and that the abstract, utterly free, rootless human being is a fiction, and not even a healthy fiction. But I have a slightly different view of why both the romantic, authentic individual, and the rootless, self-invented individual, are the wrong ways to think about identity.

Identity is a social relationship. That's why both the people who say it's innate and the people who say it's a social construct are wrong, but not completely wrong. We *are* our social relationships in many ways – not because our circumstances completely determine who we are, but because we create and become ourselves by living with others. If we have an essence, it's the essence of who we are with others. Not only the way we see ourselves reflected in their eyes, but also what we do, the way we treat each other, the person we become by the choices we make in this real, human world. And yes, that also means our sense of who we are is socially contingent. Words we use to make sense of ourselves might make no sense in another time and place, sometimes even to another generation or social group around us. But that doesn't mean our sense of self is not real. It's as real as the pain of heartbreak or the ideal of equality. It's as real as Anna's paralysis in Freud's Vienna, or the blindness that afflicts my friend's patients.

The thing we have lost, in the way identity is usually talked about, and understood, and felt, today, is that social relationships are reciprocal. How others see you is a vital part of who you are, even if you reject the way they see you and push back against it. Some identities are formed by that pushing back, refusing to accept the mainstream view of what you are. But what you can't do is force others to see you the way you want to be seen, just by demanding that they do, any more than you can force somebody to be in love with you, just because you're in love with them.

That can be a good thing. Often others can see things in us that we can't see in ourselves, things that may be just a germ of potential. That's the basis of a great educational relationship, where a teacher sees in you a person who could be a philosopher, or a competent motorcyclist, and works with you to bring that person into being.

John Berger won the 1972 Booker Prize for his novel G., in which the eponymous character is a kind of Don Giovanni at the turn of the twentieth century in Europe. That is, G. has sexual encounters with many women. It's quite a sexy book, as you might expect, but not in the way you might expect. Unlike Mozart's Don Giovanni, who keeps what we would call a spreadsheet of his conquests, G. wants to encounter each woman as the unique person that she is, breaking all sorts of social boundaries and rules as he goes. And this, not physical attraction or cunning seduction, is the secret of his appeal to them.

'The stranger who desires you and convinces you that it is truly you in all your particularity whom he desires, brings a message from all that you might be, to you as you actually are.'[363]

Seeing herself through G.'s eyes, each woman glimpses a different person she could be. Not the dutiful daughter, the good Christian, the betrothed or wife of another man, the mother of her future children, but the person in her that transcends these social categories. Berger, who had just made the non-fiction book and television series *Ways of Seeing*, was credited with first describing the 'male gaze', which shapes how women see themselves, as well as how they are seen by men. In *G.*, which he dedicates 'To Anya, and to her sisters in Women's Liberation,' there is a section about early-twentieth-century European society, called 'A Situation of Women'.[364]

'Up to then the social presence of a woman was different in kind from that of a man. A man's presence was dependent upon the promise of power which he embodied . . . The promised power may have been moral, physical, temperamental, economic, social, sexual – but its object was always exterior to the man. A man's presence suggested what he was capable of doing to you or for you.

'By contrast, a woman's presence expressed her own attitude to herself, and defined what could and could not be done to her.'[365] Their lack of power shaped the way women saw themselves in the world. It wasn't only that men looked at women as potential objects of desire, rather than as autonomous, potentially equal, fellow human beings. Women themselves acknowledged that their social role depended, largely, on how they were seen by others, and deliberately shaped the way they projected themselves into the world.

'A woman's presence was the result of herself being split in two, and of her energy being inturned. A woman was always accompanied – except when quite alone – by her own image of herself. Whilst she was walking across a room or whilst she was weeping at the death of her own father, she could not avoid envisaging herself walking or weeping. From earliest childhood she had been taught and persuaded to survey herself continually. And so she came to consider the surveyor and the surveyed within her as the two constituent yet always distinct elements of her identity as a woman.'[366]

This contrast between men and women is not the same in today's WEIRD societies. Power – economic and social, at least – is more widely distributed through society, including among women, who can have financial independence, social status and political power in their own right. A female prime minister or CEO certainly holds power over both men and women. So, in one sense, we could say that we are all men now, in that our social presence can hold some promise of power. But in another sense, we are all women, as Berger describes the situation of being constantly self-surveying. Not just because we are constantly expressing ourselves digitally, making sure our social media presence accurately projects our persona. Offline as well as online, the need to have our identity reflected back to us means we are always attending to, and anticipating, how others see us.

In his acceptance speech for the 1972 Booker Prize, at the Café Royal in London, John Berger spoke about the pleasure it gave him to have other writers respond to his book by awarding him the prize, and about the importance of literature that escapes or defies conformity and consensus, stimulates imaginative independence, and encourages people to question. 'The reason why the novel is so important is that the novel asks questions which no other literary form can ask: questions about the individual working on his own destiny; questions about the uses to which one can put a life – including one's own,' he said. 'And it poses these questions in a very private way. The novelist's voice functions like an inner voice.'[367]

I should say that this novel, which I first read as a teenager, is one that I reread every few years, always finding in it something new. I have a copy signed by John Berger when, as a star-struck student, I travelled to London to see him give a talk, and queued up to get his signature on my paperback copy. I was only vaguely aware that it had won the Booker Prize, I just knew that it spoke to me, as a reader.

In his Booker Prize acceptance speech, Berger described his next project, about the migrant workers of Europe. The prize money, £5,000, would make his research trips possible. But then he turned his attention to the source of the prize money, Booker McConnell. The company's history in the Caribbean went back over a century to the Booker brothers, on whose sugar plantation enslaved workers toiled, many of them far from the place of their birth. After receiving compensation when slavery was abolished in 1834, the brothers founded a trading and shipping business, and continued running the plantations with indentured labour.[368] One consequence of the economic history of the Caribbean is poverty that has forced hundreds of thousands of West Indians to migrate to Britain to find work. So, Berger would be researching migrant workers using money made by exploiting their ancestors, which resulted in their own situation today.

The slave trade, said Berger, has shaped the relationship between Europe and Africa, between Black and White people, not just politically and economically but psychologically. The most important single image in his novel, he said, was the statue in Livorno of four chained Moors, with distinctly Black African features, at the feet of Ferdinando de' Medici. The stone figure of Ferdinando I, considered the city's founder, stands poised, looking out to the horizon. In contrast, the four bronze figures below perch uncomfortably, hands chained behind them, looking unhappily out as they pull against their bonds, or with eyes downcast in resignation.

'Before the slave trade began, before the European dehumanised himself . . . there must have been a moment when black and white approached each other with the amazement of potential equals,' Berger's acceptance speech continued. 'The novelist is concerned with the intersection between individual and historical destiny. The historical destiny of our time is becoming clear. The oppressed are breaking through the wall of silence which was built into their minds by their oppressors. And in their struggle against exploitation and neocolonialism – but only through and by virtue of this common struggle – it is possible for the descendants of the slave and the slave-master to approach each other again with the amazed hope of potential equals.' For this reason, and not from guilt or philanthropy, or even politics, Berger said, he would give half his prize money to the British Black Panther Movement. It was about his continuing development as a writer; 'The issue is between me and the culture which has formed me.'[369]

The British Black Panther Movement, which took its name but not much else from the American Black Panther Party, was an organisation bringing together workers from the Caribbean, Africa and the Indian subcontinent, at a time of overt racism, police harassment and racial discrimination in Britain. It was a revolutionary organisation whose ultimate aim was to take the property of companies like Booker McConnell in the Caribbean and hand it over to the workers who worked there.

Berger, it is clear, did not see identity as destiny. Being Black or White in 1972 carried the history of centuries of empires and enslavement, both economically and in how mutual perception, and self-perception, are shaped by the legacy of the past. But by joining together, across colour lines, to fight for a different world, Berger saw hope of changing both reality and perception.

Appiah, writing thirty years later, also describes a process of transforming shared identities. 'There have been historical moments where we see groups contesting and transforming the meaning of their identity with seismic vigour,' he says. 'Certainly this has been a notable dimension of the grand identity movements of the twentieth century. In order to construct a life with dignity, it has seemed natural to take the collective identity and construct positive life-scripts instead.'[370] The American Black Power movement is one example he cites; the Stonewall riot, and the movement that grew out of it, is another. Both, in his account, follow similar stages.

Each movement took an old, negative script about who they are in the eyes of the oppressor, and worked together to write a new, positive script about what it means to be Black, or gay. They changed the language away from old, pejorative terms to the new names they claimed for themselves. In the course of this, they transformed their shared identity from a source of shame and hurt into something in which they can take pride, not least because they have shaped it themselves through shared struggle.

But this means that the original goal, of equal rights and equal dignity with everyone else, by virtue simply of being a person like other people, is no longer enough. Now, they demand from the rest of society respect as a member of the group that was previously despised. Gay Pride was a direct riposte to the shame that same-sex attraction was supposed to carry.

'But I think we must go on to the next step,' says Appiah, 'which is to ask whether the identities constructed in this way are ones we can be happy with in the longer run. Demanding respect for people as blacks and as gays can go along with notably rigid strictures as to how one is to be an African American or a person with same-sex desires. In a particularly fraught and emphatic way, there will be demands that are made; expectations to be met; battle lines to be drawn. It is at this point that somebody who takes autonomy seriously may worry whether we have replaced one kind of tyranny with another. We know that acts of recognition can sometimes ossify the identities that are their object.'[371]

Appiah calls this Medusa Syndrome, after the Gorgon who turned to stone whoever met her gaze. And this is my worry too.

The problem with forming yourself an identity, so both you and others recognise you in your world, is similar to the problem of having a digital doppelganger. Both are a pinning down of a person who should be ever changing and growing into somebody new. Both have in focus what can be categorised and measured about an individual, what can be observed from outside. Both reduce human beings to *what*, not *who*, we are.

We are, in today's WEIRD societies, much too concerned with how others see us. In our lonely, rootless state, often detached from the kind of communities and roots that Appiah described in his own life, we need the constant ping (literal or metaphorical) of recognition from others to tell us who we are. This is a symptom of how lost we are in the world of ever-multiplying options, leaving us at the mercy of agendas we don't even see. It's futile to think that this reflection back to us can ever be anything more than a hall of mirrors.

I'm reminded, suddenly, of the hall of mirrors at the Amsterdam Youseum, and how utterly disorientating it was to be surrounded only by reflections of myself. Completely detached from any context in the real world, any sense of scale, and from any other human beings, I felt at once vulnerable and excited, a vertigo of infinite possibilities, exhilarating and terrifying.

My criticism of our WEIRD, individualistic society is not that individuals are too free to find our own path, to write our own life story, but that we are not free enough. The illusion of infinite choice is just that, an illusion. Like the young people Peter Gray described, we have more choices about what to eat and wear, but fewer choices that let us take risks and develop personal responsibility.[372]

Technology plays a part, by offering us tailored menus that reduce our liberty to the freedom to choose between options somebody else has selected. But the technology is only the means through which we pursue our unquenchable thirst for affirmation. While our energies are inturned, in John Berger's word, we cannot fully live as free people. In order to make a life for ourselves, we need to think less about ourselves and more about living in this world.

Chapter 10: Rescuing the Person from the Personalised Century

'The moment we want to say *who* somebody is, our very vocabulary leads us astray into saying *what* he is; we get entangled in a description of qualities he necessarily shares with others like him; we begin to describe a type or a "character" in the old meaning of the word, with the result that his specific uniqueness escapes us.'[373]

HANNAH ARENDT

I said in the Introduction that the first casualty of personalisation is the person. What do I mean by that? In everyday language, we use 'person' to mean another human being, the singular form of people. But in other contexts it has more specific meanings.

For example, campaigners for animal rights sometimes claim that a highly intelligent animal like an elephant or a great ape should be treated as a person by the law. That is, the elephant should have a right to autonomy, freedom to do what it wants, beyond the protection from cruel treatment that many animals already have. I think this would be wrong, because a person, in law, has not just rights but also legal responsibilities. I can claim the protection of the law – against state oppression as well as against individual criminals – but I can also expect to stand in the dock if I am accused of a crime. If I attacked an elephant in a zoo, I would (rightly) be in court for animal cruelty. But if the elephant attacked me, it would not, because we don't expect animals to understand either moral or legal codes. We don't hold them responsible for their actions.

Legal responsibility also implies that a person has the capacity, the power and the freedom to act. Somebody who is accused of a crime but claims they weren't entirely responsible for their actions has to prove 'diminished responsibility' – for example, that they had a mental illness that made them unable to exercise normal self-control.[374] We also don't blame somebody for an action they were forced to do at gunpoint, or for failing to do something that's physically beyond them.

A person is not just legally, but morally responsible for their actions. In this sense, a child is not fully a person, because a child doesn't understand the consequences of its actions, or the full reasons why something is right or wrong. We don't protect children any less because of this – in fact, we protect them more – but we recognise that they're not yet ready to take a full place in society. That's why children don't have the vote.

Moral responsibility also implies that somebody did something on purpose, and in order to set a purpose and act on it, a person needs free will. For centuries, philosophers have argued about free will, that thing we all feel we have, but that does not fit easily into a mechanistic idea of the universe as an engine of cause and effect. Determinists argue that, if everything has a cause, nobody can freely decide to do anything. Every decision to act must have been caused by something, so the feeling that you could have chosen otherwise is an illusion. Or, if you *could* have chosen otherwise, isn't it just chance that you chose one thing and not the other? Either way, there is no room for a person to spontaneously introduce a new cause into the world of cause and effect as seen by a determinist.

My view is that the world includes a number of important things that are compatible with cause and effect, but not fully explained by a deterministic model. Life, for example, organises matter into blobs with distinct goals, often competing with one another to eat, survive and reproduce. It doesn't matter to physics which blob eats the other one, but it matters to each blob. We have consciousness, which is how I'm writing this book and you're reading it, or listening to me read it.

I'm satisfied that both life and consciousness have arisen in a material universe, through evolution, but they go beyond the capacity of physics to explain them. If I can care about whether I live or die, and be aware that I am thinking about what defines a person, it is entirely plausible to me that I can will something, put it into action, and cause something new to happen.

Most of the time, my decisions are not arbitrary or down to chance, so in that sense they do have causes. But my walking into a cold room, and seeing a fire laid and a box of matches, does not 'cause' me to light the fire, in the same way as sunlight, focused through a glass sphere on the window ledge onto a newspaper, would cause a fire to start. Reasons are not causes in the same way as physical events, because reasons also need my mind to decide that they are good enough reasons for me, here, now. They do not fully determine the outcome. Suppose I am cold, but I'm also a spy hiding in a cottage in the hills: the last thing I want to do is light a fire and send smoke up the chimney to give me away to my enemies. I'm free to refrain from acting, so if I do act, that is my free choice.

Philosophers have talked about two different kinds of free will.

St Augustine, for example, wrote about the 'Free Choice of the Will',[375] a puzzle for the early Christian philosopher as clearly it gave humans the capacity to choose between sinning and being good. Given two options – commit adultery or resist temptation – humans have free will to choose between them. But the writer Hannah Arendt identified a stronger form of free will, the freedom to initiate new things spontaneously, independent of external causes.[376] For her, the very fact of being born, entering the world as a new person, meant that beginning new things is fundamental to the human condition.

Thinking was important to Hannah Arendt, who studied and taught philosophy. She believed the thinker needed to withdraw from the world and observe it from a distance, in order to make sense of it. But it was exercising free will, and the action that followed, that to Arendt formed the individual. 'Just as thinking prepares the self for the role of spectator, willing fashions it into an "enduring I" that directs all particular acts of volition,' she wrote. 'It creates the self's *character* and therefore was sometimes understood as the . . . source of the person's specific identity.'[377]

Hannah Arendt was not an enthusiast for mass society. Writing in the late 1950s, living in New York, she saw mass production and mass consumption becoming the model for almost all of society, leaving little room for spontaneity and new beginnings. She foresaw the people around her becoming more like Skinner's rats and pigeons, with freedom only to respond to their surroundings, not to change them.

'The trouble with modern theories of behaviourism is not that they are wrong but that they could become true, that they actually are the best possible conceptualisation of certain obvious trends in modern society,' she wrote. 'It is quite conceivable that the modern age – which began with such an unprecedented and promising outburst of human activity – may end in the deadliest, most sterile passivity history has ever known.'[378]

She didn't mean, of course, that humans would stop doing anything at all. Writing during the post-war boom, she thought the mass market would keep expanding, and more and more aspects of human life would become part of that cycle of production and consumption. What she meant by activity was the kind of projects that only humans can do, and only together, where the end product is not a car, or a dress, or even a factory, but new social structures and relationships. Political organising and campaigning would be the perfect example.

To distinguish this from the labour that produces what we need to support life, and even the work that creates something new – a building, or a work of art – she called this 'action'. Much of her 'action' is speech, because that is how we generally convince other people to join with us in new projects. This, for Hannah Arendt, was central to being a person in the world. 'In acting and speaking, men show who they are, reveal actively their unique personal identities and thus make their appearance in the human world,'[379] she wrote. '*Who* somebody is or was we can know only by knowing the story of which he is himself the hero – his biography, in other words; everything else we know of him, including the work he may have produced and left behind, tells us only *what* he is or was.'[380]

Hannah Arendt herself had a remarkable life story. As a young student in 1920s Germany, she had an intense love affair with her professor, the philosopher Martin Heidegger. He went on to join the Nazi party, while she, being Jewish, had to flee across Europe after being arrested by the Gestapo. In Paris, she worked with Jewish refugee organisations for six years, helping stateless people travel to what was then the British Mandate of Palestine. During this time, she separated from her first husband and began a relationship with the man who would become her second.[381] When the Nazis occupied Paris, she was interned in a camp at Gurs in south-west France, from which she escaped. She and her second husband got to Lisbon, and from there to America, where she taught and wrote in New York until her death in 1975. Never accepting a tenured position, she was the first woman to be made a full professor at Princeton.[382]

In 1961, Arendt asked the *New Yorker* magazine to send her to Jerusalem to cover the trial of the Nazi Adolf Eichmann, who had been captured in South America. Her account of the trial used the phrase 'the banality of evil' to describe the way Eichmann had played a willing and significant role in sending millions of Jewish people to their deaths. *Eichmann in Jerusalem* was not well received by everyone, including many Jewish readers. She had reported what was said in evidence, that some Jewish council leaders had co-operated with the Nazi regime, and for this she was accused of victim-blaming. More than that, her analysis of how such a terrible evil could have happened made uncomfortable reading. The problem, she said, was not a few exceptionally evil people, but a large number of people who went along with the Reich out of fear, or reluctance to rebel, or merely out of the desire to fit in, having found a place in a system that gave them a clear role to play.[383] She also remained in contact with Martin Heidegger, philosopher, former lover and former Nazi, after the war.

Reading my brief account, I'm sure you can get a vivid idea of *who* Hannah Arendt was. You can also see how *what* she was – German, Jewish, a woman – shaped the situations in which she acted, and the roles allotted to her in those situations. But it was her actions, in circumstances that she did not choose, that both revealed and formed the person she was.

Her point on the banality of evil was that, faced with totalitarian regimes, war, even the attempted extermination of an entire people, too many of us go along with evil, not out of enthusiasm but by failing to make a moral decision to stand against it. Not enough of us show who we truly are by doing what we judge to be right.

'The connotation of courage, which we now feel to be an indispensable quality of the hero, is in fact already present in a willingness to act and speak at all, to insert one's self into the world and begin a story of one's own,' she wrote; 'courage and even boldness are already present in leaving one's private hiding place and showing who one is, in disclosing and exposing one's self. The extent of this original courage, without which action and speech and therefore . . . freedom, would not be possible at all, is not less great and may even be greater if the "hero" happens to be a coward.'384

Though she lived through extremes that I hope neither you, reader, nor I, will ever have to face, this willingness to speak out and to act is something any of us can do and thus, in Arendt's view, reveal our true selves. I agree with Arendt that people show their true selves at the moments when they use their free will, when they don't just go along with the script of what one is meant to do in such-and-such a situation. At those points, we tend to forget about how we look to others. We also tend to be unaware of what it is we're revealing – or rather, *who* it is that we are revealing ourselves to be. In Arendt's words, 'one discloses one's self without ever either knowing himself or being able to calculate beforehand whom he reveals.'[385]

This reminds me of something the psychologist William James wrote about character emerging in moments of action. 'A man's character is discernible in the mental or moral attitude in which, when it came upon him, he felt himself more deeply and intensely active and alive. At such moments there is a voice inside which speaks and says, "*This* is the real me!"'[386] Erik Erikson quotes this description as capturing the feeling of having an identity, a '*subjective sense* of an *invigorating sameness* and *continuity*',[387] which every person needs to live in the human world. As Erikson puts it, William James 'experiences it as something that "comes upon you" as a recognition, almost a surprise, rather than something strenuously "quested" after'.[388] Pointless, then, to be 'pilgrims of ourselves,' in Rome or anywhere else. We find ourselves, not by looking inwards, but by acting towards a goal beyond ourselves.

We may not be able to create this self-recognition at will, but William James describes the kind of moments that can give rise to it. They are moments when we put something in motion ourselves, hoping it will bear fruit. They include 'an element of active tension' and 'trusting outward things to perform their part' to bring the action to completeness. The tension comes from beginning something that relies upon an uncertain world to come to fruition. Erikson describes this leap of faith. 'It is an active tension (rather than a paralyzing question) – a tension which, furthermore, must create a challenge "without guaranty" rather than one dissipated in a clamor for certainty.'[389]

Erikson also points out that William James suffered an identity crisis in his late twenties, in the form of a debilitating mental breakdown. By his own account, James finally escaped from severe depression by reading an essay on free will by philosopher Charles Renouvier. James decided to live on the assumption that he could choose to think and act as he wished.[390] 'My first act of free will shall be to believe in free will,'[391] he wrote.

Sometimes, reflecting afterwards on what we did without stopping to think first, we make a discovery about ourselves and what we're capable of, for good or ill. By reflecting, by asking ourselves if that's really the person we want to be, we can also change what we do in future. This kind of reflection, surely, is exactly what Arendt thought people like Eichmann were lacking when they put aside their emotional qualms to 'do a good job' by helping to murder millions of fellow human beings.

In *The Human Condition*, Arendt suggests that the freedom to constitute oneself as an individual, by making free decisions and acting on them, can be frightening. 'The individual, fashioned by the will and aware that it could be different from what it is (character, unlike bodily appearance or talents or abilities, is not given to the self at birth) always tends to assert an "I-myself" against an indefinite "they" – all the others that I, as an individual, am *not*.' The free individual is aware that this 'me' is created by choices, freely made, for which that person must accept responsibility. 'Nothing indeed can be more frightening than . . . the "feeling" that my standing apart, isolated from everyone else, is due to free will,' Arendt continues, 'that nothing and nobody can be held responsible for it but me myself.'[392]

Whether we feel in the moment that we are being our true, undivided self, or have to reflect later to see what our actions reveal about us, let us agree with Arendt and James that we see *who* a person truly is when they act, of their own free will, rather than simply going with the flow. That means the most self-revelatory thing we can do is not to post naked selfies on Instagram, or go on *Oprah*, or publish our *Confessions*, but to stick our neck out and introduce something new into our shared social world.

It's easy to see why Arendt thinks that takes courage. It's inherently risky to do something to which you can't predict the response, let alone knowing whether you can succeed in your goal. And, for all the best efforts of super-predictors, super-computers and data analysts, anything that involves human beings is, by its very nature, unpredictable, precisely because of our unique capacity to begin new things.

'The new always happens against the overwhelming odds of statistical laws and their probability,' as Arendt puts it. In the real world, there is no 'average man'. 'The fact that man is capable of action means that the unexpected can be expected from him, that he is able to perform what is infinitely improbable. And this again is possible only because each man is unique, so that with each birth something uniquely new comes into the world.'[393]

Writing in the late 1950s, Arendt used the Hungarian Revolution of 1956 as an example of people rising up without knowing whether they would succeed. They failed, but in her eyes their tragedy showed the world that the spirit of popular revolutions, seen in Europe from 1848 onwards, was still alive.[394] This is perhaps what Kant meant when he saw hope, not in the bloody aftermath of the French Revolution, but in the hope and inspiration it ignited in onlookers.

Did the Hungarian rebels act, risking their lives, just to show others that the spirit of popular revolution was alive? Of course not. They acted to resist the oppressive Stalinist regime, with the hope, perhaps nebulous, perhaps concrete, of establishing a more democratic form of government in their country. But their actions had a broader meaning that others took from it later. In Arendt's eyes, that means their sacrifice wasn't in vain: it had purpose, albeit not the purpose the rebels had in mind.

The willingness to take this risk, to act without being able to know the outcome, or even the meaning that others will ascribe to your actions, is also the point where you reveal, indeed form, *who* you are. We can never be the sole author of our own life story, only the protagonist. We act on a stage we didn't choose, improvising lines without having any control over what other characters will say in response, or how the drama will end. Only the audience, after the play has ended, will go away and tell the narrative of who we were and what we did.

The problem is not that we care what other people think of us. We *should* care what other people think of us. We care too much about what complete strangers think of us, and we certainly need to be selective about whose judgment we trust, but those who don't care what *anyone* else thinks of them are psychopaths. The problem is that we imagine we can control what other people think of us, and that we should organise our own lives around trying to do that. We need to accept that part of being human is doing things that don't turn out as we planned or hoped, that we will be judged on those things, and that our life story will not be the one we would have written. Others will tell that story, and attach meanings to it, in ways we can't control and might not understand.

Human life is uncertain. That's not a reason to be more timid about doing things, but to embrace the opportunities life throws in our direction. By taking the initiative, going off script, making our own judgments and acting without too much self-editing, we discover who we really are and, more important, who we could become.

Trying to cling to an identity that is fixed in categories and tribes – to *what*, as opposed to *who* a person is – encourages that person, as well as everyone else, to resist change, and to resist situations and challenges that might help them to change and grow. It's like arguing for your limitations. We can't, on the other hand, abandon all sense of being a person with a past and a future, with moral responsibilities for our past actions, and commitments to our future obligations.

We all need a stable idea of who we are, not only to live in a world of other people, but to live with ourselves. We can call that an identity, or a cluster of identities made up of different social relationships. I think it is more fruitful to talk, like Hannah Arendt and William James, about character. Social relationships change, and as they change, we discover different parts of ourselves. Some of the most difficult stages of life are when we are forced to change our roles in life and discover new identities.

In 1946, living in America, Hannah Arendt wrote a short essay, 'We Refugees'. 'We lost our home, which means the familiarity of daily life. We lost our occupation, which means the confidence that we are of some use in this world. We lost our language, which means the naturalness of reactions, the simplicity of gestures, the unaffected expression of feelings. We left our relatives in the Polish ghettos and our best friends have been killed in concentration camps, and that means the rupture of our private lives.'[395]

The essay is almost unbearably raw, a description of how Jewish refugees were stripped of their places in society, and thus of their identities, first in their home countries, and then in the new countries to which they were forced to flee. In vain they tell their new friends, and themselves, 'once we were somebodies about whom people cared, we were loved by friends, and even known by landlords as paying our rent regularly'.[396]

The impact of this destruction of a shared identity is personal as well as social. Although Arendt was not an observant, religious Jew, she recognised being Jewish as part of her heritage, much as Appiah recognises the threads that braid together to form his heritage. 'To be a Jew belongs for me to the indubitable facts of my life,'[397] she said in a 1964 interview. And, asked about her work with Jewish organisations in France, she expressed something similar to Appiah's description of identities being forged anew in response to persecution: 'If one is attacked as a Jew, one must defend oneself as a Jew, not as a German or a world citizen.'[398]

In our identity-obsessed age, this is a question that confronts each of us. How can we both belong, having solidarity with others that is based in common needs and experiences and struggles, and also keep alive that universalist hope for a world where every person is equally free to create a life and build a character? How can we belong to particular, shared identities, and also be world citizens?

In *The Ethics of Identity*, Appiah tells the story of finding a handwritten note after his father's death, addressed to him and his sisters, which begins by describing their heritage – Ghanaian on his side, English on their mother's. 'But then he wrote, "Remember that you are citizens of the world." He told us that wherever we chose to live – and, as citizens of the world, we could surely choose to live anywhere that would have us – we should endeavour to leave that place "better than you found it" . . . what my father understood by citizenship . . . wasn't just a matter of belonging to a community; it was a matter of taking responsibility with that community for its destiny.'[399]

I heard something similar from Labour peer Maurice Glasman, when I interviewed him for a radio programme on the future of 'home'.[400] He was writing a book about the political meaning of home, and I asked him whether that wasn't usually associated with right-wing thinkers, and with nativist ideas that people ought to stick with the places and ways of life they inherit from their parents and grandparents.

He explained that, for him, home was not a place where one's ancestors had lived for generations. His own grandfather had walked across Europe from Ukraine, fleeing anti-Jewish pogroms in the early twentieth century, to make his home in Britain. Home, for Maurice Glasman, is not about the past. It is a place where you make a commitment in the present to take some responsibility for your neighbours, yourself, and the place you live in together. It is, he said, about 'a shared stake in the future'.

Mass society, for which Hannah Arendt had little affection, stripped human beings of their individuality, absorbing them into an interchangeable mass governed, theoretically, and later literally, by statistical rules. It made possible the dehumanisation of millions of people on battlefields and in ghettoes and concentration camps. But it also made possible social solidarity across very large groups of people with no personal ties, popular revolutions which won political rights and freedoms for individuals, and mass campaigns which changed social norms as well as laws.

The radical individual freedom promised by the personalised

century – in WEIRD societies, at least – is a fraud. It reduces individuals to passive consumers, choosing between options on a menu, instead of persons who can begin new things in the world. We still live in a mass society but, isolated from each other, we lack even the strength of coming together with shared goals and projects.

The challenge now is whether we can make that illusion of self-determination a reality, while rebuilding the frayed and battered social ties that answer the question, 'Who am I?' and much more. And that's what the last chapter will address.

Chapter 11: Courage and Hope

'Political struggle or debate is the key to good political theory. A theory is just a bunch of words – sometimes interesting to think about, but just words nevertheless – until it is tested in real life. Many a theory has delivered surprises, both positive and negative, when an attempt has been made to put it into practice.'[401]

CAROL HANISCH

The promise of freedom that the personalised century brought us, in the guise of choice, turns out to be an illusion, but the aspiration to freedom is a good one, which we should make real if we can.

The connection with others that our forebears had in the mass century, though not always in forms they had chosen, is today mostly mediated through technology, and not in forms we control. That connection is also a good thing to aim for – but not necessarily in any of these forms, past or present. Where, then, do we go from here?

The answer isn't to get rid of technology from our lives. It's not even to demand that tech companies make the data-gathering and profiling systems more transparent and more in our control, though I do think that would be a good thing. We need to turn the promise of freedom that the personalised century gives us, in the form of choice from a menu, into real freedom.

That means not just choosing from predetermined options

– whether to sin or not, as Augustine saw it – but beginning new things and setting the direction of our own lives. That also means accepting the inherent riskiness of the human world, the uncertainty of the future, the moral responsibility for setting things in motion that will have consequences you couldn't foresee, didn't intend and wouldn't have chosen.

I don't blame you if right now you're wishing I *had* just told you to throw your mobile phone in the sea and go back to sending letters written with a quill pen. Life is hard enough without swimming against the flow. Life is especially hard on your own. Cut off from the web of support we weave for each other, individuals struggle to find the necessities – food, shelter, warmth – but also to have a place and purpose that gives their lives meaning. The fellow Jewish refugees Hannah Arendt wrote about in her essay didn't just find it hard to build new lives, new homes, new jobs. Many of them lost the will to live.

I have felt for a while that, in our WEIRD societies, neither social solidarity nor our individual sense of self were looking very healthy. These are long-term trends, as described in the earlier chapters. But the experience of the Covid pandemic gave these thoughts a new urgency.

The emergency measures against the pandemic brought society's normal patterns of co-operation and interaction to a screeching halt, and left individuals heavily reliant on technology to connect with each other, and with the wider world. It was like putting the slow trends of the personalised century into fast forward. The effects on individuals were stark. Reasonable fear of a new and dangerous disease, equally reasonable fear of the long- and short-term economic effects, and the sudden rupture of all normal patterns of life, made everybody's future instantly more uncertain. Couple this with isolation from normal social contacts, and individuals had to cope with this uncertainty, and in many cases with illness and bereavement, more alone than any previous generation facing a crisis.

In the brief period of watching a contagious disease spreading across the world in early 2020, before the first wave of infection broke in Britain, I was asked on a TV current affairs programme what individuals could do. My reply was. 'Get to know your neighbours, because you're going to need each other.' I was mocked by some who felt I should have said 'stock up on pasta' or invented some public health advice, but I felt vindicated as restrictions on movement tightened, supply chains faltered, and an already creaking healthcare system staggered under the burden of illness. I did, indeed, rely on my friends and neighbours more than ever, and did what I could to support them too.

I was far from alone. Twelve million people in Britain, roughly one in six of the population, volunteered, formally or informally, during the pandemic.[402] A government call for volunteers to help the NHS got half a million responses in twenty-four hours, rising to three quarters of a million in four days. Local religious and community groups organised shopping and food supplies for people sheltering at home. Thousands of Mutual Aid groups sprang up in neighbourhoods to provide practical and emotional support, especially for those self-isolating.[403] Later, volunteers helped out at vaccination centres.

Taken by surprise, the UK Government failed to lead this early wave of social solidarity, or even to give it elbow room. In the first week of the official NHS Volunteer Responders scheme, only 20,000 tasks were given to the 750,000 volunteers.[404] Although the official headline message was that we're all in this together, government messaging quickly moved to portraying other people as the threat. Internal government communications discussed 'ramping up . . . the fear/guilt factor' and plans to 'frighten the pants of [sic] everyone'.[405]

Psychologists on the UK Government's own scientific advice team later expressed alarm and regret about how much official messaging had aimed to provoke fear and guilt to encourage compliance with restrictions, calling it 'ethically questionable' and 'dystopian'.[406] I still remember a horrific series of posters showing patients in oxygen masks, one of which bore the slogan, 'Look him in the eyes and tell him you can't work from home.' Half the country was still going out to work, I thought, and thank heavens they were.

People like me can largely work from home, but only because other people keep producing food and getting it onto supermarket shelves, keep clean water running out of my taps, gas into my boiler and electricity through my wiring, keep my internet connection live, and so on. Then there's the layer of human activities that keep *those* people at work; fuelling their vehicles, stocking their spare parts, educating their kids while they're out at work, and of course the millions working at superhuman rates in the health service itself. The people *not* working from home were the last possible people I should be blaming for the rolling tide of tragedy inflicted by the virus.

Meanwhile, research revealed a worryingly low level of self-isolation among those who had tested positive for Covid. Less than half of the people who were actually infectious self-isolated fully.[407] The reasons given for breaking self-isolation were largely practical – to go to the shops for food, or being unable to afford to take time off work. Two things that could have made a big difference to them were volunteer support and better sick pay provision. Millions of ordinary Brits were trying to deliver the former, with very uneven support from officialdom. The latter was in the hands of government, who did nothing despite repeated warnings about low self-isolation rates.

So the story of the pandemic is partly a story of people revealing who they are at a time of crisis, and rising to the challenge of a shared emergency with individual and collective altruism. Unfortunately, it's also a story of the frayed and stretched fabric of society being shredded with little regard for the long-term damage that would cause.

Education was forced online to a great extent, though children of key workers and some other families were still allowed to attend school in the UK. The long-term effects of this went beyond students falling behind academically. There is now a crisis of non-attendance: in 2022–3, one in five pupils missed 10 per cent or more of their lessons.[408] Many universities, once they had discovered that they could offer online delivery of content as 'education', were reluctant to bring students back into rooms to interact, another blow to the idea that education is a shared project of opening and developing minds, rather than a conveyor belt for facts that get turned into grades, in exchange for fees.

What was left of public space for exchanging ideas, building networks and nurturing relationships followed a similar fate. Gatherings, formal and informal, indoor and outdoor, were repeatedly forbidden and sometimes broken up by police. Digital platforms suddenly became the main way to connect, exchange ideas and make shared projects happen. Online discussions and meetings have some advantages over meetings in physical rooms. They're open to a wider geographical audience, to people with caring responsibilities or mobility issues, and they're often easier to fit into a schedule. But online meetings lack the informal and non-verbal interactions, the catching somebody's eye across an audience, or the unguarded, off-the-record chat over a drink afterwards, which can be the basis of turning an idea into a shared project. It's hard to form strong bonds of trust when all you know of somebody is a tiny face on a screen.

All these impacts are important, and we are very far from returning to a pre-pandemic state of public affairs. Especially heart-breaking to me was a reckless willingness to abandon live events, as if the audience is a dispensable element of a football match, a play or a concert. Clearly, the reason to keep crowds of spectators away was to reduce the chance for the virus to spread: on a scale of priorities in a pandemic, neither live music nor live sports rank very high. Nevertheless, it troubled me that spectators were treated as optional. Researchers have found that attending live sporting events is linked to feeling less lonely, to more life satisfaction, and more sense of purpose and meaning in life.[409] Coming together to share experiences matters to human beings.

Football matches resumed, but to empty stadiums. Quickly realising how much crowd response adds to the experience of remote listeners, broadcasters added synthetic crowd noise to live match commentary. But the absence of supporters had an impact on players as well. The usual advantage to the home team was halved.[410] 'You have to ask whether it is worth playing football without the spectators,' said Pep Guardiola, manager of Manchester City, in

March 2020. 'It doesn't make any sense to play professional football without the people, because they are the ones we do it for.'[411]

Long after gathering in crowds ceased to pose a serious health risk, especially outdoors, live events were not reinstated. The BBC went ahead with a Covid-adapted 2020 Proms season, named for the promenaders who stand in the arena of the Royal Albert Hall, enjoying world-class live music at low prices. The concerts were broadcast live as usual, but with no live audience, not even a few seated listeners scattered around the vast auditorium. I heard the first one on the radio. The orchestra reached the end of a sublime piece, and the silence that followed has wiped all memory of the music from my mind. I'm sure as many people as usual were listening remotely, if not more. We had been deprived of live performance for many months. But without an audience in the hall, the performance felt meaningless and empty. I couldn't listen to any more of the season, it was too depressing.

Any performer will tell you that the audience is as much part of creating the performance as the people on the stage. A play is, in one sense, the same night after night. But though the cast and the lighting cues are the same, the audience is different. Individuals who arrived at a venue strangers to each other, and who will depart still knowing nothing about each other, form one audience for the duration of the performance. That shared response is as much part of the collective experience as witnessing what happens on stage. By coming together in the same place and time, for the same purpose, everyone contributes to what takes place.

Watching or listening remotely, as individuals in private spaces, is a completely different activity. Even if you share your thoughts and feelings as you go by double-screening, commenting via social media or messaging your friends who are watching from their own private spaces, you are an individual, and you are receiving, not participating. Isolated individuals, even connected via the internet, remain individuals, with all the vulnerability that brings.

Conspiracy theories flourished, as people living through times that made little sense found themselves too much alone, and too much online. Sitting around a pub table, if somebody tells you that Bill Gates is injecting everyone with a 5G tracker chip disguised as a vaccine, it's easy to point out that, if Bill Gates really wanted to track us, we all carry mobile phones already. But without the stabilising influence of everyday interactions, eccentric and extreme ideas flourish, especially when frightened individuals can seek them out and form online communities with others who believe them. Everybody I know went crazy at some point during the pandemic, and I include myself in that.

We've had a foretaste of an extreme version of today's atomised, technology-mediated, hyper-personalised world. Now, as the world is still in post-pandemic flux, as long-term economic consequences roll on and new geopolitical conflicts emerge, we have an opportunity to change direction.

What I'm proposing is even harder than exhorting you to be a strong individual, to stick up for what you think is right. It's impossible to be a strong individual without others. To reclaim the person from the personalised world, we need to rebuild the social bonds that give us strength. What I'm asking for is more freedom and also more social bonds. The obvious question is – aren't these two demands in conflict? Isn't it precisely the weakening of old social bonds that led us even to the limited freedom of choice we have today? Or, put the other way, wasn't it the liberation of the individual from restrictive tradition and rigid social boundaries that left the old social bonds so weak?

Looked at historically, that is largely true. The gradual process that reorganised Western societies around the individual – free to move around, change occupation, and eventually to assert individual human rights – inevitably unravelled old social structures. That unravelling left more individuals free to make their own ways in the world, travelling to cities or to new continents to make a living. Many were forced to move, leaving behind family, shared religious practices, even their languages and habits of everyday life. The two processes have fed each other. The creation of mass societies was only possible because those people became interchangeable with one another as a workforce, or a fighting force, or as voting citizens, and therefore became easy to treat as a mass, not as unique individuals.

But remember that, before this process brought those masses of people together – in circumstances very much not of their own choosing – even the idea that individuals should have equal rights to be treated as persons, by the law and by each other, would have made no sense.

Equality in the Middle Ages happened between people of the same social status, or on Judgement Day. Even the seventeenth-century philosophers and eighteenth-century writers of new constitutions and declarations couldn't imagine the world that logically followed from their abstract principles. It took individuals prepared to think outside the norms of their time to declare that all men are created equal, or that every human being ought to be treated, morally, as an end in himself, not merely used as a means to another's end. But it took masses of people, uniting to achieve a shared goal, to make those ideas a reality for themselves and for millions of others like them, women as well as men.

'You urge without ceasing the Rights of Man,' Robespierre told the French Assembly in 1791, 'but you believe in them so little yourselves that you have sanctified slavery constitutionally.'[412] The same France in which a revolution's ideals found expression in the *Declaration of the Rights of Man* was a France made rich by colonies, and the largest was San Domingo. By 1789 its trade – mainly in sugar, grown by half a million enslaved workers[413] – was worth over £10 million, double the total of Britain's colonial trade that year.[414] *The Black Jacobins,* C. L. R. James's 1938 account of the revolution that led to that colony's independence, was both an inspiration and an education for later anti-colonialist movements.

Gathered together in large groups to work on the plantations, the enslaved Africans were able to organise themselves to revolt, and a leader emerged who would prove to be a brilliant political as well as military strategist. Toussaint, born to enslaved parents, chose his surname Louverture, or L'Ouverture – opening or beginning – as a declaration of emancipation. For ten years, he led the fight for freedom on San Domingo. Slavery was abolished in the colony in 1793, and in February 1794, a San Domingan delegation – one Black, one mixed-race and one White – asked the government in France to declare the abolition of slavery in all the colonies. The vote was carried. In 1801 Governor Toussaint Louverture published a new Constitution for San Domingo, guaranteeing equal opportunities for all regardless of colour, and the permanent abolition of slavery.

Political changes in France threatened this emancipation. In December 1801, Napoleon Bonaparte despatched a large military force to regain control of the colony. After an increasingly bloody war, Toussaint Louverture was taken captive to France, where he would die in prison.

But this was not the end of the story. In July 1802, two men swam ashore from a ship in Le Cap harbour to warn the San Domingans that Bonaparte had restored slavery in Guadeloupe.[415] Now fighting for their liberty and their lives, both Black and mixed-race islanders faced a brutal policy of extermination. C. L. R. James reports that the French general 'Rochambeau drowned so many people in the bay of Le Cap that for many a long day the people of the district would not eat fish.'[416] Finally, Rochambeau admitted defeat and on the last day of 1803, the Declaration of Independence was read out. The new state had a new name, taken from the island's inhabitants before the French colonisers arrived: Haiti.

To create an independent Haiti, it wasn't enough to have the ideas of 'liberté, égalité, fraternité' – liberty, equality and brotherhood – which to this day is the national motto of France. It took courageous humans to risk their lives by rebelling against the status quo, demanding that words be turned into reality. They looked at the French revolutionaries, and the French looked back at them, in John Berger's words, 'with the amazed hope of potential equals'.

More than two centuries later, the impoverished people of Haiti are yet to fully enjoy freedom, genuine equality, or much fraternal solidarity. But their example showed that it was possible to change things. It pushed other countries and governments towards abolishing the slave trade, and the institution of slavery, for fear of revolt by the enslaved. Like the revolutionaries in France, or in Hungary, they changed the way onlookers saw the world, even when they failed to achieve all of their own aims. The author C. L. R. James played an active role in African independence movements, including the one in Ghana in which Kwame Anthony Appiah's father Joseph was also active.

This is why I see no contradiction between the hope for real personal freedom and the hope for social solidarity. Far from being in opposition, it's impossible to have one without the other. Freedom to take risks, initiate new projects, change the world we share, needs not only practical support, but a sense of purpose and wider meaning that we can only develop in solidarity with others. That solidarity is only possible between persons, between individuals who take responsibility for themselves and for each other, who are willing to set new things in motion with no guarantee they will come to fruition. Without trust and mutual support, it's impossible to be a strong individual who takes risks. Without strong individuals willing to take risks with others, it's impossible to forge bonds of trust and build networks of solidarity.

It's pointless to feel nostalgic for the kind of social bonds that existed before the seventeenth century came along and began unpicking them all. Even if we could rewind history, we'd find them stifling and oppressive. Those were the times, remember, when most people were legally tied to the place they were born, and even the clothes you wore were limited by your social status, as well as your sex. It's equally pointless to feel nostalgic for the kind of collective experiences that characterised the mass century. Not only technology, but the social changes I have described in previous chapters, have changed the meaning we attach to shared activities, from education to political demonstrations.

We do still have social bonds. Families may be more scattered geographically, but we are still bound by loyalty and love between generations. Workplaces may be fragmented, but common goals and interests can still be the basis for mutual trust and support. Neighbours may come and go more often than in previous centuries, but they still share a neighbourhood and a common interest in the fabric of everyday life, from bins to bus routes. These bonds are important. By valuing and cultivating them, we give each other both practical support and a sense of belonging. Remember, too, that it's in relationships with others that we build character. But we also need to take on the task of building wider social bonds of solidarity to tackle the challenges we face as a society.

For an overview of where we are, let's compare today with the early years of the mass century. In the nineteenth century, the people running society had a lot of ideas about progress, and how to make it. Some of those were good ideas. We owe sanitation, science and the seeds of modern medicine to this era. Cities became more habitable and healthier. Mass education brought about mass literacy. The industrial revolution gets mixed reviews these days, but without it we would not have lighting, heating, transport, telecommunications or affordable, mass-produced goods. We might have a slightly cooler planet, but we also would not have the engineering and science that let us understand climate change and give us options to tackle it. The industrial revolution prevented mass famine, mass deaths from infectious diseases, and generally left us less at the mercy of nature's ruthless ways.

Some of the governing classes' ideas of progress were not so good. They saw the growing populations they ruled as both an asset and a challenge. Remember, this is before most people could vote for their government. The British Government saw the colonies as a place to ship the surplus population to build the Empire, instead of hanging around causing trouble at home. Empire-building, turning what began as trading colonies into directly ruled territories, consolidated an imbalance of wealth and power that persists today. Some countries worried about underpopulation and the lack of a workforce to keep the economy growing. Others worried that the lower classes were having too many babies. Eugenics, the desire to breed humans like farm animals, revealed the true evil of dehumanising the masses.

Those masses had ideas of their own about what progress should be. They organised themselves, using the strength of numbers, to push for the things they thought would leave the world a better place: decent living and working conditions, for a start, and a fair share in political power. Using the collective power of fraternity, they demanded that liberty and equality be put into practice for everyone. The mass provision of healthcare, welfare, education and housing were not just enlightened projects by the governing few who believed in progress. They were also the result of the masses pushing until their political voice was too strong to ignore, and their human needs became a political priority for democratically elected governments.

Shortly before he fled Nazi Germany for the US, Hanns Eisler wrote the music for a 1932 film by Bertolt Brecht. *Kuhle Wampe* tells the story of a family hit by unemployment, suicide and homelessness, and ends with 'The Solidarity Song'.

'Forward! And don't forget,
Where our strength lies.
In hunger and in feasting,
Forward, and don't forget – Solidarity!'

The song calls on the peoples of the Earth to unite – 'Black, White, Brown and Yellow'.

> 'Forward! And never forget
> To ask the concrete question:
> In hunger and in feasting:
> Whose tomorrow is tomorrow?
> Whose world is the world?'[417]

And now?

To start with those running our Western, democratic societies, our elected governments and others in positions of power and responsibility, there is a glaring absence of ideas, and loss of faith in progress. The positive legacy of the mass era – water, sewers, electricity, fuel, transport, healthcare, welfare and housing – is in disrepair. Nobody, in government or private companies, seems willing to take responsibility for it. Despite technological progress and economic growth, some basic social provisions are worse than a hundred years ago. The idea that the next generation should live a life of more material plenty, more freedom, wider horizons and higher expectations, has gone out of fashion. We are, in fact, encouraged to make do with less: less travel, less heating, less choice in what to buy.

Demand management is now the explicit policy of many organisations and authorities whose job used to be supplying the things we need – water, energy, transport and public services. 'Old social, economic and policy certainties such as steady growth, sustained public spending and economic stability are long gone,' said a 2014 report for local government.[418] Having given up on the idea of reliable supply that is adequate to people's needs – let alone our aspirations for a better future – the idea is to lower those aspirations and expectations instead.

If the people in charge seem incapable of taking responsibility for maintaining, let alone improving, the world, what of today's masses? Does it even make sense to think in terms of 'the masses' in our post-mass-century age? I hope I've shown that, in spite of the elevation of the self-expressing individual to the pinnacle of today's society, and the flattering sense of having one's uniqueness recognised at every moment, we do still live in a mass society.

To the automated systems that serve us, govern us, hire us and sell us goods and services, we really are just numbers. Digital technology makes it easy to position each of us in a microsegment of single figures, or a single individual, but that's not because a fellow human has a special interest in us as a potential equal in a reciprocal relationship. Even some of our interactions with each other reduce us to numbers, when we're just another share or thumbs-up, adding social value to a social media post.

But there are major differences. When the mass century was young, our ancestors were physically gathered together in large numbers. They worked together, lived together, joined mass organisations, and turned out in crowds to watch events or attend public meetings. The letter from Manchester cotton mill workers to Abraham Lincoln, supporting the blockade of the slave-owning south, was sent from a mass meeting.

Today, by contrast, we relate to issues and institutions as individuals, forming clusters that cohere and break apart unpredictably. Our allegiance to political parties, and even causes, is loose and changeable. Professional pollsters and public opinion researchers often struggle to characterise or predict what populations think or do. From the point of view of the masses – that's us – this atomisation and lack of organisation is a disadvantage. We lack a collective voice, and the aggregated power that being part of a mass can bring. As an individual, there is little I can do to change anything. I'm very much the weaker party in my dealings with companies and large organisations, as I'm reminded every time I have to wait an hour on the telephone to speak to somebody who almost certainly doesn't have the power to do what I need anyway.

Not only do we generally lack a collective voice, even on issues that affect millions of us, we also lack shared places to develop our thoughts into a coherent form. Getting together with others to work out what common ground we have, and how it might form a foundation for acting together, means bringing private thoughts and priorities into a public space. Trying to convince others, or to clarify half-formed ideas, improves and winnows individual thinking. We are, as a species, better at finding the weak spots in other people's ideas than our own. Human reason is a team sport.

Freedom of speech is sometimes discussed as the right of an individual to say anything, a form of self-expression. This is to miss the point of freedom of conscience, speech and assembly. Without them, democracy and other shared social projects are impossible. The right to assemble makes little sense as an individual, but plenty as a means to organising a group or campaign. All other forms of freedom are built on the freedom to believe, to argue, to meet with others, and to hear arguments that may change your mind. Freedom of speech is really the freedom to hear, to read and to watch what you choose, instead of allowing others to decide what ideas are fit for the masses to encounter. It's not about all ideas being equally good. On the contrary, it's about testing and challenging bad ideas in public.

In the US, freedom of belief, speech and assembly are protected by the First Amendment to the Constitution: 'Congress shall make no law respecting an establishment of religion, or prohibiting the free exercise thereof; or abridging the freedom of speech, or of the press; or the right of the people peaceably to assemble, and to petition the Government for a redress of grievances.'[419] You can see its roots in the early settlers whose religious freedom of conscience was linked to their sense that individual citizens should freely consent to be governed.

Freedom of speech and the press, and freedom of assembly, are not in the US Constitution because people have an individual right to self-expression. An early draft, written by James Madison in June 1789, read: 'The people shall not be deprived or abridged of their right to speak, to write, or to publish their sentiments; and the freedom of the press, as one of the great bulwarks of liberty, shall be inviolable.'[420] He saw clearly that the ability of the mass of people to discuss politics and read about current affairs was essential to democracy.

Technology has plenty of potential to provide new shared spaces for news, current affairs and public debate. Polarised exchanges and fractured world views reveal our worst tendencies to defend our identities at all costs, but they're not the whole picture. People do use the internet and social media to reach out and forge new connections, and to seek out information that challenges their own assumptions. We don't design the technology that shapes our communication, but we can decide how we use it.

In order to change the direction our society is sliding, with nobody apparently at the wheel, we urgently need to forge new social bonds and find new ways to organise ourselves towards common goals. Technology, as the main way we communicate and connect today, can be a tool in making this happen. But the other essential element is individuals who are prepared to take a lead, to risk rejection and failure by being the first to push against the status quo on whatever front they decide is worth fighting for.

The bad news is that such individuals are rare in today's world. In this world, it's very easy indeed to conform, because everyone can have their own unique, personalised brand of conformity. You can be one in a million, simply by making choices from your customised menus that nobody else has made. You can express your nonconformity in ways designed to acknowledge your protest and present it back to you as social validation. Want to rebel against stereotypes that tell you how to dress and behave? No problem, not only can you style yourself however you wish, you can choose an entire identity, complete with words and symbols, so you can stand out from the crowd while belonging to a tribe. Want to speak out against social injustice? There's probably a slogan for that, and maybe a flag. Sure, nobody seems sure what demands accompany it, or how to turn those demands into change, but simply to display the slogan or flag shows you're on the right side of history.

Why try to engage with people who disagree with you, and attempt to change their minds? Much easier and safer to provoke them to respond in ways that prove you were right all along. There are pre-written scripts in which you can play the public role of the person holding the views with which you feel at home. There are tribes waiting to welcome you to their warm embrace, with an entire portfolio of opinions, language and uniform, to signal to that tribe that you belong, and to others that the tribe has your back.

I asked Carol Hanisch, author of the original article published as 'The Personal Is Political', how she thought politics has changed since 1969. She said that a more apt description now would be 'the political is personal'. Hanisch was sceptical of identity as a way of understanding and organising politics.

'I don't think we gain anything by it,' she told me. 'We lose all sense of a need for unity to take on the source(s) of our oppression. Instead it invites us to escape the real world by "living in our heads". It allows us to believe that if we can change ourselves and/or just a few people, everything will be fine, which is, of course, nonsense. We're talking society-wide oppression here, not just individual attitudes. You can't "identify" yourself out of oppression, though over the centuries people sure have tried!'

Her words echoed Kwame Anthony Appiah's warning about Medusa Syndrome, the 'demands that are made; expectations to be met; battle lines to be drawn',[421] which ossify the very identities forged in struggle against the status quo. Instead of movements for liberation, they become rigid, unmoving structures of limitation, and easily slide into policing acceptable opinions and behaviour.

'This has also led to what has become known as "de-platform-ing",' Hanisch told me, 'a less political term for censorship, both of ourselves and others. It's ironic in the sense that a silencing fear of expressing oneself on political issues has been created in the name of protecting someone else's "self-expression". Nothing is so threatening to progress as fear to critique what is.'

It's not so easy to think for yourself, to 'critique what is', as a first step to making progress. If you go off script, you risk falling foul of

Appiah's 'rigid strictures', and being expelled from the tribe. You risk finding yourself alone. But you have the possibility of calling on others to join you in a new kind of tribe, where solidarity is based on common interests, on trust, on shared goals for change. A tribe where orthodoxies are there to be challenged and debated.

The good news is that, when people do step out of the chorus line and start improvising a new scene, that creates an opportunity for other individuals to respond. People who didn't think they were heroes will find that they have done something heroic, in Hannah Arendt's sense of a willingness to act and speak: to reveal in public who you are by what you do, instead of by what you declare yourself to be. The more people find themselves in social groups oriented, not inwards at affirming an identity, but outwards at changing the wider world, the more people will feel confident enough to take such risks.

If you've read this far, I have hope that you are one of the courageous people.

This is no small thing I'm asking you to do. This is a call to change the direction our society, like all WEIRD societies, is moving. We have an opportunity to turn away from the steady slide into an impersonally personalised, microsegmented world. That world is worse than mass society, because it leaves masses of people without the strength and support of being part of a mass, but also without the freedom to determine our own futures.

Take courage. Look backwards at how far we've come in such a short time. The material plenty we have, even in the post-pandemic recession, would dazzle our great-grandparents. Our capacity to save the sick and prevent illness would amaze them too. All this was the work of human minds and hands, working together.

The real freedoms we do have, to live as we wish, and to expect equal treatment no matter our sex or skin colour or nationality or religion, were hard-fought battles won by people like us. If they could change so much in a century or two, what couldn't we do?

This book is the stranger who brings a message, to all that you are, from all that you might be.

You can list your identities: the ones you were born into, the ones your life grew into, the ones you chose yourself. You could contact all the data companies and download the data that makes up their profiles of you. I'm speaking now to the part of you that eludes capture by all those labels and data points, the part of you that doesn't even recognise yet *who* you could become.

You will never know what you are capable of until you begin. You cannot know beforehand whether you will succeed. But even if you fail, you will become the person who has the courage to try.

Endnotes

Introduction

1   'The State of Podcasts and Smart Speakers', Cumulus Media Audioscape 2023, https://www.westwoodone.com/wp-content/uploads/2023/04/Cumulus-Media-2023-Audioscape_WWO.pdf.

2   Michal Kosinski et al., 'Private Traits and Attributes Are Predictable from Digital Records of Human Behavior', *PNAS*, 110, no. 15 (11 March 2013): 5802–5805, https://www.pnas.org/doi/10.1073/pnas.1218772110.

3   'Sam Smith Buys £12m London Pad after Changing Pronoun to "They" *Sun*, 14 September 2019, https://www.thesun.co.uk/tvand-showbiz/9929486/sam-smith-new-mansion-12-million-non-binary/.

Chapter 1: From Mass Production to Customised Consumption

4   Claire M. Segijn and Iris van Ooijen, 'Differences in Consumer Knowledge and Perceptions of Personalized Advertising: Comparing Online Behavioural Advertising and Synced Advertising', *Journal of Marketing Communications* 28, no. 2 (17 February 2022): 207–26, https://doi.org/10.1080/13527266.2020.1857297.

5   'Attitudes towards online advertising in the UK 2023', Statista, accessed 22 September 2023, https://www.statista.com/forecasts/997866/attitudes-towards-online-advertising-in-the-uk.

6   Samuel Crowther and Henry Ford, 'My Life and Work', Chapter IV, accessed 13 April 2023, https://www.gutenberg.org/files/7213/7213.txt

7   'Our Heritage – The Ford Motor Company Story', Ford UK, accessed 15 May 2023, https://www.ford.co.uk/experience-ford/news/our-heritage.

8   'Figure 1-1. Vehicle Ownership Rates: The United States from 1900 To . . . ', ResearchGate, accessed 13 April 2023, https://www.researchgate.net/figure/1-Vehicle-Ownership-Rates-The-United-States-from-1900-to-2000-and-15-Other-Countries_fig1_264967305.

9   '2021 Ford Edge', Preston Ford accessed 11 December 2022, https://www.prestonford.com/static/brand-ford/vehicle/2021/bro-chures/2021_Ford_Edge.pdf.

10 GBR Editorial Team, 'Ford Implemented Lean Manufacturing in Production System', *George Business Review* (blog), 14 September 2021, https://www.george-business-review.com/ford-implemented-lean-manufacturing-in-production-system/.

11 *Amy Purdy in Paralympic Rio 2016 Opening Ceremony*, YouTube, n.d., https://youtu.be/xcvYRjWdfyU.

12 Imran Amed and Achim Berg, 'The State of Fashion 2019', McKinsey & Company, accessed 5 January 2022, https://www.mckinsey.com/~/media/mckinsey/industries/retail/our%20insights/fashion%20on%20demand/the-state-of-fashion-2019.pdf.

13 El Mehdi Sehasseh et al., 'Early Middle Stone Age Personal Ornaments from Bizmoune Cave, Essaouira, Morocco', *Science Advances* 7, no. 39: eabi8620, accessed 15 December 2021, https://doi.org/10.1126/sciadv.abi8620.

14 'Those Earrings Are So Last Year – But the Reason You're Wearing Them Is Ancient', University of Arizona News, 22 September 2021, https://news.arizona.edu/story/those-earrings-are-so-last-year-reason-youre-wearing-them-ancient.

15 Jane O'Grady, 'Elizabeth Anscombe', *Guardian*, 11 January 2001,

sec. Education, https://www.theguardian.com/news/2001/jan/11/
guardianobituaries.highereducation.

16 'Who Was Wearing the Trousers', Adam Smith Institute, 28 May 2019,
https://www.adamsmith.org/blog/who-was-wearing-the-trousers.

17 'City-of-Chicago-v.-Wilson-Supreme-Court-of-Illinois-United-States.
Pdf', 1, accessed 29 November 2021, https://www.icj.org/wp-content/
uploads/2012/07/City-of-Chicago-v.-Wilson-Supreme-Court-of-
Illinois-United-States.pdf.

18 What (Not) to Wear: Fashion and the Law: Sumptuous Origins:
Colonial Massachusetts, Harvard Law School, accessed November
2023, https://exhibits.law.harvard.edu/colonial-massachusetts.

19 'Sumptuary Law of 1651 Massachusetts Bay Colony and the Fairbanks
Family', Fairbanks History, accessed 2 December 2021, https://www.
fairbankshistory.com/colonial-history/sumptuary-law-of-1651-mas-
sachusetts-bay-colony-and-the-fairbanks-family.

20 Sara Semic, 'Influencers Are Now Buying Virtual Clothes They Will
Wear On IG But Never Touch IRL', *ELLE*, 2 July 2019, https://www.
elle.com/uk/fashion/a28166986/digital-fashion-dressing-virtually/.

21 George Horace Gallup, 'An objective method for determining reader
interest in the content of a newspaper', University of Iowa, https://
iro.uiowa.edu/esploro/outputs/doctoral/An-objective-method-for-
determining-reader/9983776807002771

22 Henry Durant, 'Public Opinion, Polls and Foreign Policy',
British Journal of Sociology 6, no. 2 (June 1955): 151, https://doi.
org/10.2307/587480.

23 M. Roodhouse, '"Fish-and-Chip Intelligence": Henry Durant and
the British Institute of Public Opinion, 1936–63', *Twentieth Century
British History* 24, no. 2 (1 June 2013): 225, https://doi.org/10.1093/
tcbh/hws012.

24   John B. Watson, 'Psychology as the Behaviorist Views It', n.d., 248.

25   'Pioneer — Edward Bernays', Museum of Public Relations, accessed 14 December 2021, https://www.prmuseum.org/pioneer-edward-bernays.

26   Iris Mostegel, 'Edward Bernays: The Original Influencer', *History Today*, accessed 14 December 2021, https://www.historytoday.com/miscellanies/original-influencer.

27   Christopher A. Summers, Robert W. Smith and Rebecca Walker Reczek, 'An Audience of One: Behaviorally Targeted Ads as Implied Social Labels', *Journal of Consumer Research* 43, no. 1 (1 June 2016): 156–78, https://doi.org/10.1093/jcr/ucw012.

Chapter 2: From Mass Media to Your Own Channel

28   'A Short History of the BBC', BBC News, accessed 26 January 2022, http://news.bbc.co.uk/1/hi/entertainment/1231593.stm.

29   Larry Rohter, 'An Unlikely Trendsetter Made Earphones a Way of Life', *New York Times*, 17 December 2005, sec. World, https://www.nytimes.com/2005/12/17/world/americas/an-unlikely-trendsetter-made-earphones-a-way-of-life.html.

30   'Speech Opening Conference on Information Technology', Margaret Thatcher Foundation, accessed 9 March 2022, https://www.margaretthatcher.org/document/105067.

31   '10 Things You Didn't Know about Ceefax', BBC News, 23 September 2009, http://news.bbc.co.uk/1/hi/magazine/8260196.stm.

32   James Curran and Jean Seaton, *Power Without Responsibility*, 8th ed. (Routledge, 2018), 354.

33   'The Netflix Prize: Crowdsourcing to Improve DVD Recommendations', *Digital Innovation and Transformation* (blog), accessed 11 December

2022, https://d3.harvard.edu/platform-digit/submission/the-netflix-prize-crowdsourcing-to-improve-dvd-recommendations/.

34 Xavier Amatriain and Justin Basilico, 'Netflix recommendations: Beyond the 5 stars', Netflix Technology Blog, 6 April 2012, https://netflixtechblog.com/netflix-recommendations-beyond-the-5-stars-part-1-55838468f429

35 Ibid.

36 Curran and Seaton, *Power Without Responsibility*, 49.

37 Ibid., 50.

38 Daniel J. Robinson, *The Measure of Democracy: Polling, Market Research, and Public Life, 1930-1945* (Toronto: University of Toronto Press, 1999), 15.

39 'Channel 4's £1m Diversity in Advertising Award Challenges Industry to Authentically Represent UK BAME Cultures in Advertising', Channel 4, accessed 13 April 2023, https://www.channel4.com/press/news/channel-4s-ps1m-diversity-advertising-award-challenges-industry-authentically-represent.

40 Andrew Kersley, 'AA/WARC: Record £30bn UK ad spend in 2021, but growth for publishers to slow this year', Press Gazette, 28 January 2022, https://pressgazette.co.uk/uk-ad-spend-2021/.

41 'Worldwide Ad Spending 2021: A Year for the Record Books', Insider Intelligence, accessed 17 May 2023, https://www.insiderintelligence.com/content/worldwide-ad-spending-2021-year-record-books.

42 Robert Hart, 'Facebook Loses Daily Active Users For The First Time – Here's Where They're Going', *Forbes*, accessed 30 November 2022, https://www.forbes.com/sites/roberthart/2022/02/03/facebook-loses-daily-active-users-for-the-first-time--heres-where-theyre-going/.

43 'State of Mobile 2022', Data.Ai, accessed 30 November 2022, https://www.data.ai/en/go/state-of-mobile-2022.

44 Ibid.

45 Sara Atske, 'Teens, Social Media and Technology 2022', *Pew Research Center: Internet, Science & Tech* (blog), 10 August 2022, https://www.pewresearch.org/internet/2022/08/10/teens-social-media-and-technology-2022/.

46 'Digital News Report 2022', Reuters Institute, accessed 15 June 2022, https://reutersinstitute.politics.ox.ac.uk/sites/default/files/2022-06/Digital_News-Report_2022.pdf.

47 Ibid.

48 Chris Bail, *Breaking the Social Media Prism* (Princeton University Press, 2021), 119.

49 Ibid., 93.

50 Ibid., 93–95.

51 Ibid., 100.

52 Ibid., 14.

53 'Most Facebook and Twitter Users' Online Networks Contain a Mix of People with a Variety of Political Beliefs', *Pew Research Center: Internet, Science & Tech* (blog), accessed 12 December 2022, https://www.pewresearch.org/internet/2016/10/25/the-political-environment-on-social-media/pi_2016-10-25_politics-and-social-media_1-02/.

54 Elizabeth Dubois and Grant Blank, 'The Echo Chamber Is Overstated: The Moderating Effect of Political Interest and Diverse Media', *Information, Communication & Society* 21, no. 5 (4 May 2018): 741, https://doi.org/10.1080/1369118X.2018.1428656.

55 Seth Flaxman, Sharad Goel and Justin M. Rao, 'Filter Bubbles, Echo Chambers, and Online News Consumption', *Public Opinion Quarterly* 80, no. S1 (1 January 2016): 298–320, https://doi.org/10.1093/poq/nfw006.

56 Sara Atske, 'Connection, Creativity and Drama: Teen Life on Social

Media in 2022', *Pew Research Center: Internet, Science & Tech* (blog), 16 November 2022, https://www.pewresearch.org/internet/2022/11/16/connection-creativity-and-drama-teen-life-on-social-media-in-2022/.

57 'About The Orwell Foundation' The Orwell Foundation, https://www.orwellfoundation.com/the-orwell-foundation/about/about-the-orwell-foundation-2/

58 'George Orwell at the BBC', The Orwell Foundation, accessed 29 November 2022, https://www.orwellfoundation.com/the-orwell-foundation/news-events/news-events/news/george-orwell-at-the-bbc/.

Chapter 3: From Mass Movements to the Personal Is Political

59 Andrew Therriault, *Data and Democracy* (O'Reilly Media, Inc., 2016), 45.

60 Ibid.

61 Sasha Issenberg, 'How President Obama's Campaign Used Big Data to Rally Voters', *MIT Technology Review*, December 2012, https://www.hereweare.com/read_a_more_perfect_union.

62 Ed Pilkington and Amanda Michel, 'Obama, Facebook and the Power of Friendship: The 2012 Data Election', *Guardian*, 17 February 2012, sec. US news, https://www.theguardian.com/world/2012/feb/17/obama-digital-data-machine-facebook-election.

63 Hannes Grassegger and Mikael Krogerus, 'The Data That Turned The World Upside Down', *Vice*, 2017, https://www.vice.com/en/article/mg9vvn/how-our-likes-helped-trump-win.

64 Benjamin Kidd, *Social Evolution* (London: Methuen, 1920), 11.

65 Gustave Le Bon, *The Psychology of Crowds* (Sparkling Books, 2020), 9.

66 Ibid.

67 Ibid., 10.

68 Ibid., 10.

69   '1842 and 1848 Chartist Petitions', UK Parliament, accessed 17 August 2021, https://www.parliament.uk/about/living-heritage/transform-ingsociety/electionsvoting/chartists/case-study/the-right-to-vote/the-chartists-and-birmingham/1842-and-1848-chartist-petitions/.

70   Chartist Ancestors, 'Monster Meeting on Kennington Common, 10 April 1848', *Chartist Ancestors* (blog), 18 August 2017, https://www.chartistancestors.co.uk/monster-meeting-kennington-common-10-april-1848/.

71   Francis Galton, 'Eugenics: It's Definition, Scope, and Aims', *American Journal of Sociology* 10, no. 1 (July 1904): 1–25.

72   Ibid.

73   Kevin Duong, 'Universal Suffrage as Decolonization', *American Political Science Review* 115, no. 2 (May 2021): 412–28, https://doi.org/10.1017/S0003055420000994.

74   Elizabeth Clery et al., *British Social Attitudes 36*, 2019, 191, https://natcen.ac.uk/publications/british-social-attitudes-36

75   Ibid.

76   'Aaron Copland + Fanfare for the Common Man', Kennedy Center, accessed 4 April 2023, https://www.kennedy-center.org/education/resources-for-educators/classroom-resources/media-and-interactives/media/music/aaron-copland--fanfare-for-the-common-man/.

77   'Henry Wallace - Free World Association Dinner Address (Transcript-Audio)', American Rhetoric, accessed 12 August 2021, https://www.americanrhetoric.com/speeches/henrywallacefreeworldassoc.htm.

78   Preston Valien, 'The Brotherhood of Sleeping Car Porters', *Phylon* 1, no. 3 (1940): 233, https://www.jstor.org/stable/271990.

79   'The Scottsboro Boys', National Museum of African American History and Culture, 15 March 2017, https://nmaahc.si.edu/blog/scottsboro-boys.

80 '(1941) Executive Order 8802', Black Past, 24 January 2007, https://www.blackpast.org/african-american-history/executive-order-8802-1941-2/.

81 'Rustin, Bayard', The Martin Luther King, Jr. Research and Education Institute, 31 May 2017, https://kinginstitute.stanford.edu/encyclope-dia/rustin-bayard.

82 'Martin Luther King I Have a Dream Speech', American Rhetoric, accessed 2 August 2021, https://www.americanrhetoric.com/speeches/mlkihaveadream.htm.

83 Ibid.

84 Tom Hayden, *The Long Sixties: From 1960 to Barack Obama* (Paradigm, 2009), 29.

85 The Democracy Journal - The Port Huron Statement: A Manifesto Reconsidered', Tom Hayden, accessed 18 September 2021, https://www.latimes.com/opinion/la-xpm-2012-may-06-la-oe-hayden-port-huron-statement-20120506-story.html.

86 Ibid.

87 '1848 Chartist Petition', UK Parliament, accessed 18 September 2021, https://www.parliament.uk/about/living-heritage/transform-ingsociety/electionsvoting/newport-rising/1839-newp-ris/univ-suff-pet-1848/.

88 'The Port Huron Statement, Written by Tom Hayden', American Progress, accessed 29 January 2024, https://images2.americanpro-gress.org/campus/email/PortHuronStatement.pdf.

89 Ibid., 4.

90 Ibid., 5.

91 Ibid., 5.

92 Ibid., 5.

93 Ibid., 9.

94 Carol Hanisch, 'The Personal Is Political: The Women's Liberation Movement classic with a new explanatory introduction', Carole Hanisch, http://www.carolhanisch.org/CHwritings/PIP.html.

95 Ibid.

96 Ibid.

97 Florence Binard, 'The British Women's Liberation Movement in the 1970s: Redefining the Personal and the Political', *Revue Française de Civilisation Britannique. French Journal of British Studies* 22, no. hors-série (13 December 2017): sec. 17, https://doi.org/10.4000/rfcb.1688.

98 'Thatcherism, Trade Unionism and All That', Adam Smith Institute, accessed 23 August 2021, https://www.adamsmith.org/blog/politics-government/thatcherism-trade-unionism-and-all-that.

99 Kate Kerrow and Rebecca Mordan, *Out of the Darkness, Greenham Voices 1981–2000* (The History Press, 2021), 37.

100 Bobby Duffy et al., 'Divided Britain? Polarisation and fragmentation trends in the UK', The Policy Institute, King's College London, 2019, https://www.kcl.ac.uk/policy-institute/assets/divided-britain.pdf.

101 Kerrow and Mordan, *Out of the Darkness,* 78.

102 'London Dockers' Strike', Spartacus Educational, accessed 8 September 2021, https://spartacus-educational.com/TUdockers.htm.

103 Brian Towers, 'Running the Gauntlet: British Trade Unions under Thatcher, 1979–1988', *Industrial & Labor Relations Review* 42, no. 2 (January 1989): 163–188.

104 Ibid.

105 'Trade Union Membership UK 1995-2021 Statistical Bulletin', n.d.

106 'Public attitudes towards multiple sector strike action', Ipsos, January 2023, accessed 9 October 2023, https://www.ipsos.com/sites/default/files/ct/news/documents/2023-01/Ipsos%20UK%20January%202023%20Strikes%20polling_300123_v2.pdf.

107 *Cambridge City Councillor Kevin Price. Resignation Speech at Full Council - 22.10.2020*, YouTube, 2020, https://www.youtube.com/watch?v=WTDwli-H-f4.

108 Anna Fitzpatrick, 'Kevin Price: Life "should be about more than just existing"', Varsity Online, accessed 9 September 2021, http://www.varsity.co.uk/news/11731.

109 Emma Yeomans, 'Cambridge Row over "Transphobic" Porter Kevin Price', *The Times*, accessed 9 September 2021, https://www.thetimes.co.uk/article/cambridge-row-over-transphobic-porter-kevin-price-8bb7px950.

110 Gaby Vides, 'Cambridge University Liberal Association Condemn Transphobia in Cambridge Labour Party', Varsity Online, accessed 9 September 2021, https://www.varsity.co.uk/news/19916.

111 'Agenda for Council on Thursday, 22nd October, 2020, 6.00 Pm', Cambridge City Council, 22 October 2020, https://democracy.cambridge.gov.uk/ieListDocuments.aspx?CId=116&MId=3778&Ver=4.

112 'Attitudes toward Elected Officials, Voting and the State', *Pew Research Center's Global Attitudes Project* (blog), 27 February 2020, https://www.pewresearch.org/global/2020/02/27/attitudes-toward-elected-officials-voting-and-the-state/.

113 Ben Norton, 'Into the Streets May First by Aaron Copland (Sheet Music, Audio, MIDI File, History)', *Ben Norton* (blog), 10 September 2020, https://bennorton.com/into-the-streets-may-first-aaron-copland/.

114 Duffy et al., 'Divided Britain? Polarisation and fragmentation trends in the UK', 44.

115 Ibid., 60.

116 Ibid., 60.

117 Ibid., 58.

118 Issenberg, 'How President Obama's campaign used big data to rally voters'.

119 Ibid.

Chapter 4: From File Cards to Profiles: Technology

120 Ian Hacking, *The Emergence of Probability* (Cambridge University Press, 2007), 45.

121 F.N. David, *Games, Gods and Gambling* (Charles Griffin, 1962), 102.

122 David, *Games, Gods and Gambling*, 99–100.

123 Hacking, *The Emergence of Probability*, 106.

124 John Graunt, *Natural and Political Observations Mentioned in a Following Index, and Made upon the Bills of Mortality by John Graunt . . . ; with Reference to the Government, Religion, Trade, Growth, Ayre, Diseases, and the Several Changes of the Said City*, 1662, 18, http://name.umdl.umich.edu/A41827.0001.001.

125 'Life Expectancy Calculator', Office for National Statistics, accessed 27 April 2022, https://www.ons.gov.uk/peoplepopulationandcommunity/healthandsocialcare/healthandlifeexpectancies/articles/lifeexpectancycalculator/2019-06-07.

126 Ian Hacking, *The Emergence of Probability*, Cambridge University Press, 2006, 105.

127 Justus Nipperdey, *Johann Peter Süssmilch: From Divine Law to Human Intervention*, Population-E 66, no. 3–4 (2011): 620.

128 Immanuel Kant et al., *Toward Perpetual Peace and Other Writings on Politics, Peace, and History* (Yale University Press, 2006), sec. IUH 8:15,

129 Adolphe Quetelet, *A Treatise on Man and the Development of His Faculties* (Cambridge University Press, 2013), 5.

130 Ian Hacking, *The Taming of Chance* (Cambridge University Press, 2008), 114.

131 Quetelet, *A Treatise on Man and the Development of His Faculties*, 7.

132 Ibid.

133 Francis Galton, 'Anthropometric Laboratory', 1884, Galton.org, accessed 12 October 2022, https://galton.org/essays/1880-1889/galton-1884-anthro-lab.pdf.

134 'Museum History', Computer History Museum, accessed 5 July 2022, https://www.computerhistory.org/chmhistory/.

135 Herman Hollerith, 'Patent 395781', United States Census Bureau, accessed 5 July 2022, https://www.census.gov/history/pdf/hollerith_patent_01081889.pdf.

136 Census History Staff US Census Bureau, 'The Hollerith Machine', United States Census Bureau, accessed 5 July 2022, https://www.census.gov/history/www/innovations/technology/the_hollerith_tabulator.html.

137 'Social Security History', Social Security Administration, accessed 5 July 2022, https://www.ssa.gov/history/fdrsignstate.html.

138 https://www.ibm.com/content/dam/connectedassets-adobe-cms/worldwide-content/arc/cf/ul/g/83/1d/Sunday_News.jpg

139 'IBM Highlights 1885–1969', IBM, accessed 6 July 2022, https://www.ibm.com/ibm/history/documents/pdf/1885-1969.pdf.

140 'Eugenics: Its Origin and Development (1883 - Present)', Genome.gov, accessed 12 July 2022, https://www.genome.gov/about-genomics/educational-resources/timelines/eugenics.

141 'Eugenics in the Nordic Countries', Nordics Info, accessed 12 July 2022, https://nordics.info/show/artikel/eugenics-in-the-nordic-countries/.

142 'Pain and Redemption of WWII Interned Japanese-Americans', BBC

News, 18 February 2012, sec. Magazine, https://www.bbc.com/news/magazine-17080392.

143 'Justice pour René Carmille (1906)', *La Jaune et la Rouge* (blog), 13 March 2020, https://www.lajauneetlarouge.com/justice-pour-rene-carmille-1906/.

144 'Les services statistiques français pendant l'Occupation', Wikisource, accessed 13 July 2022, https://fr.wikisource.org/wiki/Les_services_statistiques_fran%C3%A7ais_pendant_l%E2%80%99Occupation.

145 'Des apparences à la réalité: le "fichier juif". Rapport de la commission présidée par René Rémond au Premier ministre : Mise au point par Robert Carmille', Wikisource, accessed 13 July 2022, https://fr.wikisource.org/wiki/Des_apparences_%C3%A0_la_r%C3%A9alit%C3%A9_:_le_%22fichier_juif%22._Rapport_de_la_commission_pr%C3%A9sid%C3%A9e_par_Ren%C3%A9_R%C3%A9mond_au_Premier_ministre_:_Mise_au_point_par_Robert_Carmille.

146 Ibid.

147 'A History of Hacking - IEEE', The Institute, 4 April 2015, https://web.archive.org/web/20150404142154/http://theinstitute.ieee.org/technology-focus/technology-history/a-history-of-hacking.

148 'A Brief History of the Internet', *Internet Society* (blog), accessed 7 October 2022, https://www.internetsociety.org/internet/history-internet/brief-history-internet/.

149 'The Internet Comes From Behind', Computer History Museum, accessed 7 October 2022, https://www.computerhistory.org/revolution/networking/19/378.

150 Gordon E. Moore, 'Cramming More Components onto Integrated Circuits, Reprinted from Electronics, Volume 38, Number 8, April 19, 1965, Pp.114 Ff.', *IEEE Solid-State Circuits Society Newsletter* 11, no. 3 (September 2006): 33–35, https://doi.org/10.1109/N-SSC.2006.4785860.

151 Ibid., 33.

152 Barnaby J. Feder, 'COMPUTER MARKET STRONG IN BRITAIN', *New York Times*, 10 October 1983, sec. Business, https://www.nytimes. com/1983/10/10/business/computer-market-strong-in-britain.html.

153 'Sinclair launches the ZX Spectrum', Centre for Computing History, accessed 7 October 2022, http://www.computinghistory.org.uk/ det/43781/Sinclair%20ZX%20Spectrum%20Launched.

154 'Sinclair ZX Spectrum 48k', Centre for Computing History, accessed 7 October 2022, http://www.computinghistory.org.uk/det/424/Sinclair-ZX-Spectrum-48k/.

155 'Sinclair launches the ZX Spectrum'.

156 'Tim Berners-Lee publishes the first ever website', Centre for Computing History, accessed 7 October 2022, http://www.comput-inghistory.org.uk/det/65289/Tim%20Berners-Lee%20publishes%20 the%20first%20ever%20website.

157 'Technology in the American Household', *Pew Research Center - U.S. Politics & Policy* (blog), 24 May 1994, 4, https://www.pewresearch.org/ politics/1994/05/24/technology-in-the-american-household/.

158 Kate Youde, 'Broadband: The First Decade', *Independent,* 21 December 2021, https://www.independent.co.uk/tech/broadband-the-first-decade-1929515.html.

159 'Internet/Broadband Fact Sheet', *Pew Research Center: Internet, Science & Tech* (blog), accessed 10 October 2022, https://www. pewresearch.org/internet/fact-sheet/internet-broadband/.

160 'A Legacy of Innovation: Timeline of Motorola history since 1928', Motorola Solutions, accessed 11 October 2022, https://www.motoro-lasolutions.com/en_us/about/history/timeline.html.

161 'Breakthrough for Mobile Telephony', Ericsson, 29 August 2016,

https://www.ericsson.com/en/about-us/history/products/mobile-telephony/breakthrough-for-mobile-telephony.

162 'The Invention of Mobile Phones', Science Museum, accessed 11 October 2022, https://www.sciencemuseum.org.uk/objects-and-stories/invention-mobile-phones.

163 Tom Ough, 'The Surprising Ways Cellphones Have Changed Our Lives', BBC, accessed 3 April 2023, https://www.bbc.com/future/article/20230331-the-surprising-way-cellphones-changed-our-lives.

164 'Motorola DynaTAC 8000x', Mobile Phone Museum, accessed 11 October 2022, https://www.mobilephonemuseum.com/phone-detail/dynatac-8000x.

165 'Breakthrough for Mobile Telephony'.

166 Cornelia Connolly et al., 'The Many Twists and Turns in the History of the Mobile Phone', RTÉ, 3 September 2021, https://www.rte.ie/brainstorm/2021/0901/1244162-mobile-phone-history-networks-standards/.

167 https://www.sciencemuseum.org.uk/objects-and-stories/invention-mobile-phones

168 John B. Horrigan, 'Consumption of Information Goods and Services in the U.S.', *Pew Research Center: Internet, Science & Tech* (blog), 23 November 2003, https://www.pewresearch.org/internet/2003/11/23/consumption-of-information-goods-and-services-in-the-u-s/.

169 'UK Households: Ownership of Mobile Telephones 1996-2018', Statista, accessed 11 October 2022, https://www.statista.com/statistics/289167/mobile-phone-penetration-in-the-uk/.

170 'History', NTT, accessed 11 October 2022, https://group.ntt/en/group/history/.

171 'Macworld Expo Keynote Live Update: Introducing the iPhone',

Macworld, accessed 12 October 2022, https://www.macworld.com/article/183052/liveupdate-15.html.

172 'Live from Macworld 2007: Steve Jobs Keynote', Engadget, accessed 11 October 2022, https://www.engadget.com/2007-01-09-live-from-macworld-2007-steve-jobs-keynote.html.

173 Facebook et al., 'The 10 Best Symbian Phones Ever', *PCMag UK*, 24 January 2013, https://uk.pcmag.com/mobile-phones/58161/the-10-best-symbian-phones-ever.

174 'RIM: A Brief History from Budgie to BlackBerry 10', Engadget, accessed 11 October 2022, https://www.engadget.com/2013-01-28-rim-a-brief-history-from-budgie-to-blackberry-10.html.

175 'Macworld Expo Keynote Live Update'.

176 Matt Hamblen, 'Android Smartphone Sales Leap to Second Place in 2010, Gartner Says', Computerworld, 9 February 2011, https://www.computerworld.com/article/2512940/android-smartphone-sales-leap-to-second-place-in-2010--gartner-says.html.

177 Facebook et al., 'The 10 Best Symbian Phones Ever'.

178 'Mobile Operating System Market Share Worldwide', StatCounter Global Stats, accessed 13 October 2022, https://gs.statcounter.com/os-market-share/mobile/worldwide.

179 'The Mobile Economy Europe 2022', *The Mobile Economy* (blog), accessed 13 October 2022, https://www.gsma.com/mobileeconomy/europe/.

180 'The Mobile Economy North America 2022', *The Mobile Economy* (blog), accessed 13 October 2022, https://www.gsma.com/mobileeconomy/northamerica/.https://www.gsma.com/mobileeconomy/northamerica/

181 'The Mobile Economy Sub-Saharan Africa 2021', GSM Association, 2021, accessed 13 October 2022, https://www.gsma.com/

mobileeconomy/wp-content/uploads/2021/09/GSMA_ME_SSA_2021_English_Web_Singles.pdf.

182 'The Mobile Economy 2022', GSM Association, 2022, 6, accessed 13 October 2022, https://www.gsma.com/mobileeconomy/wp-content/uploads/2022/02/280222-The-Mobile-Economy-2022.pdf.

183 'A Tale of Deleted Cities', CHM, 21 September 2016, https://computer-history.org/blog/a-tale-of-deleted-cities/.

184 Isabella Steger, 'GeoCities Japan Is Finally Shutting Down', Quartz, 1 October 2018, https://qz.com/1408120/yahoo-japan-is-shutting-down-its-website-hosting-service-geocities.

185 Timothy Stenovec, 'Myspace's Biggest Moments: Memories Of A Fallen Social Network', HuffPost, 29 June 2011, https://www.huffpost.com/entry/myspace-history-timeline_n_887059.

186 Thanasis Papadopoulos, 'A Timeline Of MySpace, Your Ex-Social Medium', *Medium* (blog), 5 May 2021, https://thanasispapadopoulos.medium.com/a-timeline-of-myspace-your-ex-social-medium-b2e29fa9aa18.

187 Nick Seaver, *Computing Taste: Algorithms and the Makers of Music Recommendation* (Chicago: University of Chicago Press, 2022), 65.

Chapter 5: From Necessity to Freedom: Choice

188 Unzipped Staff, 'What's Your Levi's® Style?', Levi Strauss & Co., 23 August 2017, https://www.levistrauss.com/2017/08/23/whats-levis-style/.

189 Kate Loveman, 'Buying Books in Restoration London', in *Samuel Pepys and His Books: Reading, Newsgathering, and Sociability, 1660–1703*, ed. Kate Loveman (Oxford University Press, 2015), chap. 6, https://doi.org/10.1093/acprof:oso/9780198732686.003.0007.

190 'Currency Converter: 1270–2017', National Archives, accessed 11 May 2023, https://www.nationalarchives.gov.uk/currency-converter/.

191 Casey Hall, 'Chinese Authorities Slap Comedy Firm with $2 Million Fine after Military Joke', Reuters, 17 May 2023, sec. Media & Telecom, https://www.reuters.com/business/media-telecom/chinese-slaps-comedy-firm-with-2-mln-fine-after-military-joke-2023-05-17/.

192 Confirmed by author's correspondence with the Draper's Company.

193 'The East India Company', The History of London, accessed 26 October 2022, https://www.thehistoryoflondon.co.uk/the-east-india-company/.

194 'The Royal African Company - Supplying Slaves to Jamestown', US National Park Service, accessed 24 October 2022, https://www.nps.gov/jame/learn/historyculture/the-royal-african-company-supplying-slaves-to-jamestown.htm.

195 'London's Original and All-Inspiring Coffee House', Atlas Obscura, accessed 13 December 2022, http://www.atlasobscura.com/places/londons-original-all-inspiring-coffee-house.

196 Christopher Hill, *The Century of Revolution* (Van Nostrand Reinhold (UK) Co. Ltd, 1988), 176.

197 'History: 1537–1799', Honourable Artillery Company, accessed 24 October 2022, https://hac.org.uk/where-we-come-from/history/1537-1799.

198 David Flintham, *Civil War London: A Military History of London under Charles I and Oliver Cromwell* (Helion & Company Ltd, 2017).

199 Ibid., 39.

200 'The Debates', The Putney Debates of 1647, accessed 24 October 2022, http://www.theputneydebates.co.uk/the-debates/.

201 'This Day in History – the Mayflower Departs England', *Anglican*

Mainstream (blog), 16 September 2020, https://anglicanmainstream. org/this-day-in-history-the-mayflower-departs-england/.

202 'What Was the Mayflower Compact? It's Meaning and Significance', Christianity.com, accessed 24 October 2022, https://www.christianity. com/church/church-history/timeline/1601-1700/the-magnificent-mayflower-compact-11630074.html.

203 Anthony Kenny, *A New History of Western Philosophy* (Oxford University Press, 2010), 712–16.

204 'The Debates'.

205 Hill, *The Century of Revolution*, 127.

206 Valbona Muzaka, 'Stealing the Common from the Goose: The Emergence of Farmers' Rights and Their Implementation in India and Brazil', *Journal of Agrarian Change* 21, no. 2 (2021): 356–76, https://doi.org/10.1111/joac.12398.

207 'The East India Company', The History of London, accessed 26 October 2022. https://www.thehistoryoflondon.co.uk/the-east-india-company/

208 Ibid.

209 'History', The East India Company, accessed 26 October 2022, https:// www.theeastindiacompany.com/pages/history.

210 Ibid.

211 David S. Landes, *The Unbound Prometheus: Technological Change and Industrial Development in Western Europe from 1750 to the Present* (Cambridge University Press, 1972), 85.

212 'Abe Lincoln and the "sublime Heroism" of British Workers', BBC News, 17 January 2013, sec. World, https://www.bbc.com/news/world-21057494.

213 Ibid.

214 Ibid.

215 Ibid.

216 Letter from Manchester mill workers to Abraham Lincoln (image), accessed 31 October 2022, https://static.guim.co.uk/sys-images/Guardian/Pix/pictures/2013/1/31/1359635861273/Mill-workers-001.jpg.

217 Lincoln's Letter to the Working-Men of Manchester, England', American Civil War Society (UK), accessed 31 October 2022, https://acws.co.uk/archives-misc-lincoln_letter.

218 'Overview of the Thirteenth Amendment', Constitution Annotated/Congress.Gov/Library of Congress, accessed 31 October 2022, https://constitution.congress.gov/browse/essay/amdt13-1/ALDE_00000991/.

219 Lynn Downey, 'A Short History of Denim', Levi Strauss & Co., accessed 26 October 2022, https://levistrauss.com/wp-content/uploads/2014/01/A-Short-History-of-Denim2.pdf.

220 A Brief History of Lee Jeans', Heddels, 30 July 2012, https://www.heddels.com/2012/07/brief-history-of-lee-jeans/.

221 Brandon Tensley, 'How Denim Became a Political Symbol of the 1960s', Smithsonian Magazine, accessed 26 October 2022, https://www.smithsonianmag.com/arts-culture/denim-political-symbol-1960s-180976241/.

222 Downey, 'A Short History of Denim'.

223 'Our History', Wrangler, accessed 13 December 2022, https://www.wrangler.com/history.html.

224 Downey, 'A Short History of Denim'.

225 Wouter Munnichs, 'Levi's Denim Art Contest Catalogue From 1974', *Long John* (blog), 20 March 2014, https://www.craftcouncil.org/post/levis-denim-art-contest

226 'Levi's Denim Art Contest', American Craft Council, accessed 26 October 2022, https://www.craftcouncil.org/post/levis-denim-art-contest.

227 REZPECT, 'Eye On Design: Studded Levi's Denim Jacket By Billy Shire', *The Worley Gig*, 15 March 2017, https://worleygig.com/2017/03/15/eye-on-design-studded-levis-denim-jacket-by-billy-shire/.

228 John O. Koehler, *Stasi: The Untold Story of the East German Secret Police*, Chapter 1 extract, New York Times archive, accessed 27 October 2022, https://archive.nytimes.com/www.nytimes.com/books/first/k/koehler-stasi.html.

229 Simon Fanshawe, *The Power of Difference* (Kogan Page, 2022), 2–3.

230 'Grimsby Dock Tower', *Discover North East Lincolnshire* (blog), accessed 31 October 2022, https://www.discovernortheastlincolnshire.co.uk/things-to-do/heritage-and-history/landmarks/dock-tower/.

Chapter 6: Pilgrims of Ourselves: Identity

231 'Pantheon, Rome: History and Description. Dome and Oculus', *ArcheoRoma* (blog), 19 October 2016, https://www.archeoroma.org/sites/pantheon/.

232 'Project "From Tourist to Pilgrim"', *Pantheon Rome* (blog), accessed 18 January 2024, https://www.pantheonroma.com/who-we-are/.

233 Sigmund Freud, *Civilization and Its Discontents*, Penguin Classics (Penguin, 2002).

234 Ibid., 70.

235 'Mithras Sol Invictus: An Initiate's Guide', Tastes of History, 15 February 2021, https://www.tastesofhistory.co.uk/post/mithras-sol-invictus-an-initiate-s-guide.

236 Tom Holland, *Dominion* (Abacus, 2020), 80.

237 Larry Siedentop, *Inventing the Individual* (Penguin, 2014), 119.

238 Charles Taylor, *Sources of the Self : The Making of the Modern Identity* (Harvard University Press, 1989), 129.

239 Remi Jedwab et al., 'The Economic Impact of the Black Death',

Institute for International Economic Policy, August 2020, 34, accessed 18 August 2022, https://www2.gwu.edu/~iiep/assets/docs/papers/2020WP/JedwabIIEP2020-14.pdf.

240 Jedwab et al., 'The Economic Impact of the Black Death', 5.

241 Niccolò Machiavelli, *The Prince, Selections from The Discourses, and Other Writings*, ed. John Plamenatz, trans. Allan H. Gilbert (Fontana/Collins, 1975), 7.

242 James Hall, *The Self Portrait a Cultural History* (Thames & Hudson, 2015), 72.

243 Ibid., 96.

244 Francesco Borghesi, 'Chronology', in *Pico Della Mirandola: Oration on the Dignity of Man: A New Translation and Commentary*, ed. Francesco Borghesi et al. (Cambridge University Press, 2012), 37–44, https://doi.org/10.1017/CBO9781139059565.004.

245 Ibid.,

246 Pier Cesare Bori, 'The Historical and Biographical Background of the Oration', in *Pico Della Mirandola: Oration on the Dignity of Man*, 10–36, https://doi.org/10.1017/CBO9781139059565.003.

247 Taylor, *Sources of the Self*, 200.

248 Francesco Borghesi et al., eds., 'Text', in *Pico Della Mirandola: Oration on the Dignity of Man*, secs 19–22, https://doi.org/10.1017/CBO9781139059565.008.

249 Machiavelli, *The Prince, Selections from The Discourses, and Other Writings*, 152.

250 Ibid. 200.

251 Ibid., 207.

252 4343 Luther in Rome', Augnet, accessed 14 December 2022, http://augnet.org/en/history/people/4341-martin-luther/4343-luther-in-rome/.

253 Joshua J. Mark, 'Luther's Speech at the Diet of Worms', World History Encyclopedia, accessed 14 December 2022, https://www.worldhistory. org/article/1900/luthers-speech-at-the-diet-of-worms/.

254 Kenan Malik, *The Quest for a Moral Compass* (Atlantic Books, 2014), 167–75.

255 Ibid., 175.

256 Machiavelli, *The Prince, Selections from The Discourses, and Other Writings*, 221–22.

257 Harvey Mansfield, 'Niccolò Machiavelli', Britannica, accessed 24 August 2022, https://www.britannica.com/biography/Niccolo-Machiavelli.

258 Borghesi et al., 'Text', sec. 118.

259 Taylor, *Sources of the Self*, 183.

260 Ibid., 179.

261 Anthony Kenny, *A New History of Western Philosophy* (Oxford University Press, 2010), 538–39.

262 Taylor, *Sources of the Self*, 172.

263 Christopher Bertram, 'Jean Jacques Rousseau', in *The Stanford Encyclopedia of Philosophy*, ed. Edward N. Zalta, Winter 2020 (Metaphysics Research Lab, Stanford University, 2020), https://plato. stanford.edu/archives/win2020/entries/rousseau/.

264 Tim Black, 'Autonomy and the Birth of Authenticity', in *From Self to Selfie* (Palgrave Macmillan, 2019), 119.

265 Jean-Jacques Rousseau and Ligaran, *Les Confessions* (Ligaran Éditions, 2019), 7.

266 Immanuel Kant et al., *Toward Perpetual Peace and Other Writings on Politics, Peace, and History* (Yale University Press, 2006), 269.

267 Hannah Arendt, *Lectures on Kant's Political Philosophy* (University of Chicago Press, 1982), 48.

268 Michael Forster, 'Johann Gottfried von Herder', in *The Stanford Encyclopedia of Philosophy*, ed. Edward N. Zalta, Summer 2022 (Metaphysics Research Lab, Stanford University, 2022), https://plato.stanford.edu/archives/sum2022/entries/herder/.

269 Jerrold Seigel, *The Idea of the Self* (Cambridge University Press, 2005), 433.

270 'Timeline of the Napoleonic Era - 1806', Napoleon & Empire, accessed 14 December 2022, https://www.napoleon-empire.net/en/chronology/chronology-1806.php.

271 Seigel, *The Idea of the Self*, 457–58.

272 John Stuart Mill, *On Liberty, Utilitarianism, and Other Essays* (Oxford University Press, 2015), 44.

273 Morton Hunt, *The Story of Psychology* (Anchor, 2007), 142.

274 Eric R. Kandel, *The Age of Insight* (Random House, 2012), 45.

275 Hunt, *The Story of Psychology*, 186–92.

276 Kandel, *The Age of Insight*, 63.

277 Paolo Benanti, *Homo Faber: The Techno-Human Condition* (Edizioni Dehoniane Bologna, 2018).

278 'B. F. Skinner', Harvard University, accessed 14 December 2022, https://psychology.fas.harvard.edu/people/b-f-skinner.

279 B.F. Skinner, '"Superstition" in the Pigeon', 1948, Hanover Education, 273, accessed 14 December 2022, https://psych.hanover.edu/classes/Learning/papers/Skinner%20Superstion%20(1948%20orig).pdf.

280 B.F. Skinner, Beyond Freedom and Dignity (Penguin Books, 1971), 10, accessed 14 December 2022, https://ia601303.us.archive.org/27/items/Beyond_Freedom_and_Dignity/Beyond%20Freedom%20%26%20Dignity%20-%20Skinner.pdf.

281 Ibid., 194.

282 Benanti, *Homo Faber*.

Chapter 7: The Narcissist in the Hall of Mirrors

283 Christopher Lasch, *Culture of Narcissism* (Norton, 1991), 91.

284 'Sigmund Freud', Institute of Psychoanalysis, accessed 14 December 2022, https://psychoanalysis.org.uk/our-authors-and-theorists/sigmund-freud.

285 Helmut Dahmer, 'Adorno's View of Psychoanalysis', *Thesis Eleven* 111, no. 1 (1 August 2012): 97–109, https://doi.org/10.1177/0725513612453422.

286 Merve Emre, *What's Your Type?* (William Collins, 2018), 126.

287 T.W. Adorno and Else Frenkel-Brunswik, *The Authoritarian Personality* (Harper & Brothers, 1950), 971.

288 Ibid., 975–76.

289 Erik H. Erikson, *Identity Youth and Crisis* (W.W. Norton, 1968), 23.

290 Ibid., 22.

291 Lasch, *Culture of Narcissism*, 25.

292 Ibid., 10.

293 Ibid., 84.

294 Alisha Ebrahimji, Dakin Andone and Amir Vera, 'Buffalo shooting victims: "Hero" guard and a teacher who was a "pillar of the community" are among 10 killed', CNN, accessed 18 May 2022, https://www.cnn.com/2022/05/15/us/buffalo-shooting-victims-what-we-know/index.html.

295 Lasch, *Culture of Narcissism*, 47.

296 'Celebrations | Liturgical Life I Pantheon Roma - Official Site', *Pantheon Rome* (blog), accessed 21 July 2022, https://pantheonroma.org/home-eng/.

Chapter 8: Right for You and Only You

297 *At Experian, we know what makes you one of a kind – Goths advert*, YouTube, 2019, https://youtu.be/6jX1kBjQga8.

298 Sara Atske, '2. Americans Concerned, Feel Lack of Control over Personal Data Collected by Both Companies and the Government', *Pew Research Center: Internet, Science & Tech* (blog), 15 November 2019, https://www.pewresearch.org/internet/2019/11/15/americans-concerned-feel-lack-of-control-over-personal-data-collected-by-both-companies-and-the-government/.

299 Sara Atske, '3. Public Knowledge and Experiences with Data-Driven Ads', *Pew Research Center: Internet, Science & Tech* (blog), 15 November 2019, https://www.pewresearch.org/internet/2019/11/15/public-knowledge-and-experiences-with-data-driven-ads/.

300 *Good Ideas Deserve to Be Found: A (Slightly) Life-Changing Story*, Meta, 2022, https://www.facebook.com/Meta/videos/1459847961114516/.

301 'A Brief History of Experian: Our Story', Experian, 2013, accessed 21 November 2022, https://www.experianplc.com/media/1323/8151-exp-experian-history-book_abridged_final.pdf.

302 Josh Lauer, *Creditworthy: A History of Consumer Surveillance and Financial Identity in America* (Columbia University Press, 2017), 29.

303 Ibid., 32–33.

304 'A Brief History of Experian', 9.

305 Ibid., 18.

306 'Sears History - 1890s', Sears Archives, accessed 21 November 2022, http://searsarchives.com/history/history1890s.htm.

307 Daniel J. Robinson, *The Measure of Democracy: Polling, Market Research, and Public Life, 1930–1945* (University of Toronto Press, 1999), 73–74, https://doi.org/10.3138/9781442681712.

308 Ibid., 85–86.

309 'R. L. Polk & Co.', Company-Histories.com, accessed 21 November 2022, https://www.company-histories.com/R-L-Polk-Co-Company-History.html.

310   Lauer, *Creditworthy*, 254.

311   According to Polk, 'R.L. Polk & Co: All in the Family', *America According to Polk* (blog), 12 April 2019, https://www.accordingtopolk.com/2019/04/12/r-l-polk-co-family/.

312   'R. L. Polk & Co.'.

313   Lauer, *Creditworthy*, 254.

314   'PRIZM® Premier', Claritas LLC, accessed 23 November 2022, https://claritas.com/prizm-premier/.

315   'PRIZM Postal Code Lookup Demo', Environics Analytics, Default, accessed 23 November 2022, https://environicsanalytics.com/en-ca/PRIZM.

316   Richard James Webber, Tim Butler and Trevor Phillips, 'Adoption of Geodemographic and Ethno-Cultural Taxonomies for Analysing Big Data', *Big Data and Society*, no. January-June 2015 (2015): 1–16, https://doi.org/0.1177/2053951715583914.

317   Nadia El-Yaouti, 'California Woman Wrongly Arrested, Sues for $2.5M', Law Commentary, 28 February 2022, https://www.lawcommentary.com/articles/california-woman-wrongly-arrested-sues-for-25m.

318   'User-friendly behavioral biometrics authentication', TypingDNA, https://www.typingdna.com/behavioral-biometrics.html

319   'Get a Personal Identity Number (Personnummer) in Sweden', Swedish Immigration, accessed 21 November 2022, https://www.swedishimmigration.se/all-topics/working-in-sweden/how-to-get-a-personal-identity-number-personnummer-in-sweden/.

320   Timandra Harkness, 'The History of the Data Economy: Part III: The new Kings and Queens of data', Significance Vol 18 issue 5, October 2021, 16-19: https://doi.org/10.1111/1740-9713.01567.

321   Megan Graham Elias Jennifer, 'How Google's $150 Billion Advertising Business Works', CNBC, accessed 15 November 2022, https://www.

cnbc.com/2021/05/18/how-does-google-make-money-advertising-business-breakdown-.html.

322   Muhammad Ali et al., 'Discrimination through Optimization: How Facebook's Ad Delivery Can Lead to Skewed Outcomes', *Proceedings of the ACM on Human-Computer Interaction* 3, no. CSCW (7 November 2019): 4, https://doi.org/10.1145/3359301.

323   Amit Datta, Michael Carl Tschantz and Anupam Datta, 'Automated Experiments on Ad Privacy Settings: A Tale of Opacity, Choice, and Discrimination', Cornell University, (arXiv, 16 March 2015), https://doi.org/10.48550/arXiv.1408.6491.

324   'Personalized Advertising', Google, accessed 18 November 2022, https://support.google.com/adspolicy/answer/143465#552.

325   Sheri Hughes, 'Half of Job Adverts Unconsciously Biased towards Male Applicants', theHRDIRECTOR, 21 March 2021, https://www.thehrdirector.com/business-news/diversity-and-equality-inclusion/half-of-job-adverts-unconsciously-biased-towards-male-applicants/.

326   Ali et al., 'Discrimination through Optimization', 12.

327   Ibid., 14.

328   '"Half the Money I Spend on Advertising Is Wasted; the Trouble Is I Don't Know Which Half." | B2B Marketing', accessed 18 November 2022, https://www.b2bmarketing.net/en-gb/resources/blog/half-money-i-spend-advertising-wasted-trouble-i-dont-know-which-half.

329   Matthew Wall, 'Location tech and mobile map way out to better business', BBC News, 16 June 2014, sec. Business, https://www.bbc.com/news/business-27781078.

330   Ibid.

331   Celina Burnett and Colin Strong, *One Not Everyone* (MRS/IPA/Marketing Society, 2017), 88.

332 B.J. Fogg, *Persuasive Technology Using Computers to Change What We Think and Do* (Morgan Kaufmann Publishers, 2003), 216–17.

333 Miguel Helft, 'The Class That Built Apps, and Fortunes', *New York Times*, 7 May 2011, sec. Technology, https://www.nytimes.com/2011/05/08/technology/08class.html.

334 Fogg, *Persuasive Technology Using Computers to Change What We Think and Do*, 243.

335 Ibid., 193.

336 Ibid., 189.

337 'Change Your Habits and Life with Pavlok', Pavlok, accessed 28 November 2022, https://pavlok.com/.

Chapter 9: It's Not the Technology – it's Us

338 Kwame Athony Appiah, *The Ethics of Identity* (Princeton University Press, 2005), 23.

339 Joseph Henrich, Steven J. Heine and Ara Norenzayan, 'The Weirdest People in the World?' *Behavioral and Brain Sciences* 33, no. 2–3 (June 2010): 71, https://doi.org/10.1017/S0140525X0999152X.

340 Ibid.

341 Ibid.

342 Ibid.

343 Ola Svenson, 'Are We All Less Risky and More Skillful than Our Fellow Drivers?' *Acta Psychologica* 47, no. 2 (February 1981): 146, https://doi.org/10.1016/0001-6918(81)90005-6.

344 John F. Finch and Robert B. Cialdini, 'Another Indirect Tactic of (Self-) Image Management: Boosting', *Personality and Social Psychology Bulletin* 15, no. 2 (1989): 222–32, https://doi.org/10.1177/0146167289152009.

345 Emily Terlizzi and Benjamin Zablotsky, 'Mental Health Treatment

Among Adults: United States, 2019', NCHS Data Brief 380 (September 2020): 1–8, https://pubmed.ncbi.nlm.nih.gov/33054921/.

346 'Mental Health of Children and Young People in England, 2020: Wave 1 Follow up to the 2017 Survey', NHS Digital, accessed 18 December 2022, https://digital.nhs.uk/data-and-information/publications/statistical/mental-health-of-children-and-young-people-in-england/2020-wave-1-follow-up.

347 Jean Twenge, 'Teens Have Less Face Time with Their Friends – and Are Lonelier than Ever', The Conversation, 20 March 2019, http://theconversation.com/teens-have-less-face-time-with-their-friends-and-are-lonelier-than-ever-113240.

348 'Many Teens Say They're Constantly Online – but They're No Less Likely to Socialize with Their Friends Offline', *Pew Research Center* (blog), accessed 14 March 2023, https://www.pewresearch.org/fact-tank/2018/11/28/teens-who-are-constantly-online-are-just-as-likely-to-socialize-with-their-friends-offline/.

349 Helen Dodd et al., 'Children's Play and Independent Mobility in 2020: Results from the British Children's Play Survey', *International Journal of Environmental Research and Public Health* 18 (20 April 2021): 4334, https://doi.org/10.3390/ijerph18084334.

350 Peter Gray, David F. Lancy, and David F. Bjorklund, 'Decline in Independent Activity as a Cause of Decline in Children's Mental Wellbeing: Summary of the Evidence', *Journal of Pediatrics* 260 (23 February 2023): 2, https://doi.org/10.1016/j.jpeds.2023.02.004.

351 Amy Orben et al., 'Windows of Developmental Sensitivity to Social Media', *Nature Communications* 13, no. 1 (28 March 2022): 1649, https://doi.org/10.1038/s41467-022-29296-3.

352 Amy Orben, Tobias Dienlin and Andrew K. Przybylski, 'Social Media's Enduring Effect on Adolescent Life Satisfaction', *Proceedings of the*

National Academy of Sciences 116, no. 21 (21 May 2019): 10226–28, https://doi.org/10.1073/pnas.1902058116.

353 Jeffrey A. Hall, Michael W. Kearney and Chong Xing, 'Two Tests of Social Displacement through Social Media Use', *Information, Communication & Society* 22, no. 10 (24 August 2019): 1396–1413, https://doi.org/10.1080/1369118X.2018.1430162.

354 Bail, *Breaking the Social Media Prism*, 89.

355 Ibid., 39.

356 Ibid., 10.

357 Gerry Greenstone, 'The History of Bloodletting', British Columbia Medical Journal 52, no. 1 (January/February 2010): 12–14, accessed 2 December 2022, https://bcmj.org/premise/history-bloodletting.

358 Appiah, *The Ethics of Identity*, 214.

359 Ibid., 20.

360 Ibid., 156.

361 Ibid., 19.

362 Peter De Vries, *Reuben, Reuben* (Little, Brown, 1964), 242.

363 John Berger, *G.* (Dell Publishing, 1973), 146.

364 Ibid., 164.

365 Ibid.

366 Ibid., 164–65.

367 '"I Have to Turn the Prize against Itself": John Berger's 1972 Booker Prize Speech in Full', The Booker Prizes, accessed 3 December 2022, https://thebookerprizes.com/the-booker-library/features/i-have-to-turn-the-prize-against-itself-john-bergers-1972-booker-prize.

368 'The Booker Brothers and Enslavement', The Booker Prizes, accessed 17 March 2023, https://thebookerprizes.com/the-booker-brothers-and-enslavement.

369 '"I Have to Turn the Prize against Itself"'.

370 Appiah, *The Ethics of Identity*, 109.

371 Ibid., 110.

372 Gray, Lancy and Bjorklund, 'Decline in Independent Activity as a Cause of Decline in Children's Mental Wellbeing', 2.

Chapter 10: Rescuing the Person from the Personalised Century

373 Hannah Arendt, *The Human Condition* (University of Chicago Press, 1998), 181.

374 'Coroners and Justice Act 2009', Legislation.gov.uk, accessed 14 March 2023, https://www.legislation.gov.uk/ukpga/2009/25/part/2/chapter/1/crossheading/partial-defence-to-murder-diminished-responsibility.

375 Hannah Arendt, *The Life of the Mind* (Harcourt, 1978), 86.

376 Ibid., 20.

377 Ibid., 195–96.

378 Arendt, *The Human Condition*, 322.

379 Ibid., 179.

380 Ibid., 186.

381 Maurizio Passerin d'Entreves, 'Hannah Arendt', in *The Stanford Encyclopedia of Philosophy*, ed. Edward N. Zalta and Uri Nodelman, Fall 2022 (Metaphysics Research Lab, Stanford University, 2022), https://plato.stanford.edu/archives/fall2022/entries/arendt/.

382 'About Hannah Arendt at Bard College', Bard College, accessed 6 December 2022, https://hac.bard.edu/about/hannaharendt/.

383 'The Trial of Hannah Arendt', National Endowment for the Humanities, accessed 6 December 2022, https://www.neh.gov/humanities/2014/marchapril/feature/the-trial-hannah-arendt.

384 Arendt, *The Human Condition*, 186–87.

385 Ibid., 192.

386 Erikson, *Identity Youth and Crisis*, 19.

387 Ibid.

388 Ibid., 20.

389 Ibid. 20.

390 Hunt, *The Story of Psychology*, 163–64.

391 Ibid., 164.

392 Arendt, *The Life of the Mind*, 195–96.

393 Arendt, *The Human Condition*, 178.

394 Ibid., 217.

395 Hannah Arendt, 'We Refugees', in *Altogether Elsewehere* (Faber & Faber, 1994), 110.

396 Ibid., 115.

397 *Hannah Arendt - What Remains? Language Remains. (w/ English Subtitles)*, YouTube, 2016, https://www.youtube.com/watch?v=Ma5RqKdUQ8Q.

398 Ibid.

399 Appiah, *The Ethics of Identity*, 213.

400 'BBC Radio 4 - FutureProofing, Home', BBC, accessed 6 December 2022, https://www.bbc.co.uk/programmes/m00045b3.

Chapter 11: Courage and Hope

401 Hanisch, 'The Personal Is Political'.

402 Ben Leo, 'Over 12million Brits Have Volunteered during Coronavirus Pandemic', *Sun*, 28 February 2021, https://www.thesun.co.uk/news/14186257/12million-brits-volunteered-coronavirus/.

403 Guanlan Mao et al., 'What Have We Learned about COVID-19 Volunteering in the UK? A Rapid Review of the Literature', *BMC Public Health* 21, no. 1 (28 July 2021): 1470, https://doi.org/10.1186/s12889-021-11390-8.

404 Ibid.

405 The Lockdown Files Team, 'Matt Hancock's Plan to "Frighten the Pants off Everyone" about Covid', *Telegraph*, 4 March 2023, https://www.telegraph.co.uk/news/2023/03/04/project-fear-covid-lockdown-files-matt-hancock-whatsapp/.

406 Gordon Rayner, 'Use of Fear to Control Behaviour in Covid Crisis Was "Totalitarian", Admit Scientists', *Telegraph*, 14 May 2021, https://www.telegraph.co.uk/news/2021/05/14/scientists-admit-totalitarian-use-fear-control-behaviour-covid/.

407 Mao et al., 'What Have We Learned about COVID-19 Volunteering in the UK?'

408 'School Absences: The School Picking Kids up from Home to Boost Attendance', BBC News, 26 September 2023, sec. Family & Education, https://www.bbc.com/news/education-66917186.

409 Helen Keyes et al., 'Attending Live Sporting Events Predicts Subjective Wellbeing and Reduces Loneliness', *Frontiers in Public Health* 10 (2023), https://www.frontiersin.org/articles/10.3389/fpubh.2022.989706.

410 Dane McCarrick et al., 'Home Advantage during the COVID-19 Pandemic: Analyses of European Football Leagues', *Psychology of Sport and Exercise* 56 (1 September 2021): 8, https://doi.org/10.1016/j.psychsport.2021.102013.

411 Paul Wilson, 'Pep Guardiola: It Is Inevitable English Games Will Go behind Closed Doors', *Guardian*, 10 March 2020, sec. Football, https://www.theguardian.com/football/2020/mar/10/pep-guardiola-it-is-inevitable-english-games-will-go-behind-closed-doors.

412 C.L.R. James, *The Black Jacobins* (Penguin Random House, 2022), 65.

413 Ibid., 47.

414 Ibid., 43.

415 Ibid., 291–92.

416 Ibid., 306.

417 *Ernst Busch- Solidaritätslied (1931) w/ English Lyrics*, YouTube 2019, https://www.youtube.com/watch?v=FTMfqwDXUN4.

418 Anna Randle and Henry Kippin, 'Managing Demand: Building Future Public Services', RSA (2020), 15, accessed 10 December 2022, https://www.thersa.org/globalassets/pdfs/reports/rsa_managing-demand_revision4.pdf.

419 'Constitution of the United States: First Amendment', Constitution Annotated/Congress.Gov/Library of Congress, accessed 28 October 2022, https://constitution.congress.gov/constitution/amendment-1/.

420 'Historical Background on Free Speech Clause', Constitution Annotated/Congress.Gov/Library of Congress, accessed 28 October 2022, https://constitution.congress.gov/browse/essay/amdt1-7-1/ALDE_00013537/.

421 Appiah, *The Ethics of Identity*, 110.

Bibliography

Adorno, T. W., and Else Frenkel-Brunswik. *The Authoritarian Personality*. Harper, 1950.

Appiah, Kwame Athony. *The Ethics of Identity*. Princeton University Press, 2005.

Arendt, Hannah. *The Human Condition*. University of Chicago Press, 1998.

———. *The Life of the Mind*. Harcourt, 1978.

———. 'We Refugees'. In *Altogether Elsewehere*. Faber & Faber, 1994.

Bail, Chris. *Breaking the Social Media Prism*. Princeton University Press, 2021.

Bauman, Zygmunt. *Identity*. Polity, 2004.

Benanti, Paolo. *Homo Faber: The Techno-Human Condition*. Edizioni Dehoniane Bologna, 2018.

Berger, John. *G.* Dell Publishing, 1973.

Black, Edwin. *IBM and the Holocaust*. Little, Brown, 2001.

Blastland, Michael and Spiegelhalter, David. *The Norm Chronicles*. Profile, 2013.

Bon, Gustave le. *Psychology of Crowds*. Sparkling Books, 2020.

Book, Dan. *How Our Days Became Numbered*. University of Chicago Press, 2015.

Carey, John. *The Intellectuals and the Masses*. Faber & Faber, 1992.

Curran, James, and Jean Seaton. *Power Without Responsibility*. 8th ed. Routledge, 2018.

David, F.N. *Games, Gods and Gambling*. Charles Griffin, 1962.

De Vries, Peter. *Reuben, Reuben*. Little, Brown, 1964.

Descartes, René. *Meditations on First Philosophy*. Cambridge University Press, 2017.

Du Bois, W.E.B. (William Edward Burghardt). *The Souls of Black Folk*. Dover, 1994.

Emre, Merve. *What's Your Type?* William Collins, 2018.

Erikson, Erik H. *Identity Youth and Crisis*. W.W. Norton, 1968.

Fanshawe, Simon. *The Power of Difference*. Kogan Page, 2022.

Fogg, B.J. *Persuasive Technology: Using Computers to Change What We Think and Do*. Morgan Kaufmann Publishers, 2003.

Freud, Sigmund. *Civilization and Its Discontents*. Penguin Classics. Penguin, 2002.

Fukuyama, Francis. *Identity*. Profile, 2018.

Hacking, Ian. *The Emergence of Probability*. Cambridge University Press, 2007.

———. *The Taming of Chance*. Cambridge University Press, 2008.

Hall, James. *The Self-Portrait*. Thames & Hudson, 2014.

Hayden, Tom. *The Long Sixties: From 1960 to Barack Obama*. Paradigm, 2009.

Heartfield, James. *The 'Death of the Subject' Explained*. Sheffield Hallam University Press, 2002.

Hill, Christopher. *The Century of Revolution*. Van Nostrand Reinhold (UK) Co. Ltd, 1988.

Hobbes, Thomas. *Leviathan*. Penguin, 1985.

Holland, Tom. *Dominion*. Abacus, 2020.

Hunt, Morton. *The Story of Psychology*. Anchor, 2007.

James, C.L.R. *The Black Jacobins*. Penguin Random House, 2022.

Kandel, Eric R. *The Age of Insight*. Random House, 2012.

Kant, Immanuel. *Toward Perpetual Peace and Other Writings on Politics, Peace, and History.* Yale University Press, 2006.

Kennedy, Angus, and Panton, James. *From Self to Selfie.* Palgrave Macmillan, 2019.

Kenny, Anthony. *A New History of Western Philosophy.* Oxford University Press, 2010.

Kerrow, Kate, and Mordan, Rebecca. *Out of the Darkness, Greenham Voices 1981–2000.* The History Press, 2021.

Kidd, Benjamin. *Social Evolution.* Methuen, 1920.

Koopman, Colin. *How We Became Our Data.* University of Chicago Press, 2019.

Korsgaard, Christine M. *Creating the Kingdom of Ends.* 1st ed. Cambridge University Press, 1996.

———. *Self-Constitution: Agency, Identity, and Integrity.* Oxford University Press, 2009.

Landes, David S. *The Unbound Prometheus: Technological Change and Industrial Development in Western Europe from 1750 to the Present.* Cambridge University Press, 1972.

Lasch, Christopher. *Culture of Narcissism.* Norton, 1991.

Lauer, Josh. *Creditworthy: A History of Consumer Surveillance and Financial Identity in America.* Columbia University Press, 2017.

Lepore, Jill. *If Then.* John Murray, 2020.

Machiavelli, Niccolò. *The Prince, Selections from The Discourses, and Other Writings.* Edited by John Plamenatz. Translated by Allan H. Gilbert. Fontana/Collins, 1975.

Malik, Kenan. *The Quest for a Moral Compass.* Atlantic Books, 2014.

Mill, John Stuart. *On Liberty, Utilitarianism, and Other Essays.* Oxford University Press, 2015.

Quetelet, Adolphe. *A Treatise on Man and the Development of His Faculties*. Cambridge University Press, 2013.

Robinson, Daniel J. *The Measure of Democracy: Polling, Market Research, and Public Life, 1930–1945*. University of Toronto Press, 1999.

Rousseau, Jean-Jacques. *Les Confessions*. Ligaran Éditions, 2019.

Seaver, Nick. *Computing Taste: Algorithms and the Makers of Music Recommendation*. University of Chicago Press, 2022.

Seigel, Jerrold. *The Idea of the Self*. Cambridge University Press, 2005.

Siedentop, Larry. *Inventing the Individual*. Penguin, 2014.

Storr, Will. *Selfie*. Picador, 2017.

Taylor, Charles. *Sources of the Self : The Making of the Modern Identity*. Harvard University Press, 1989.

Therriault, Andrew. *Data and Democracy*. O'Reilly Media, Inc., 2016.

Thompson, E.P. *The Making of the English Working Class*. Penguin, 1991.

Turkle, Sherry. *Alone Together: Why We Expect More from Technology and Less from Each Other*. Basic Books, 2017.

Acknowledgements

My friends, especially Sandra Lawrence, for support of all kinds over the years it's taken me to get this book from first inkling to ink on a page. Likewise, my family - including Tipple and Pickle the terriers of course.

Will Francis and the team at Janklow Nesbit. My editors, and everyone at HQ stories, especially Zoe Berville, Marleigh Price, Emily Kiel and Jamie Groves.

All the people I spoke to, those who are named in the text, but also many whose words didn't make it directly onto the page, but whose insights inform what's there. Notably, I'd like to thank Michael Veale, Henriette Cramer, Linda Stone, Morgan Ames, Andrew Willshire, Carl Miller - and some people who spoke to me off the record whose names I won't put into print!

Organisers, speakers and fellow attendees at several years of The Academy, where I germinated the seeds that grew into this book, and presented an early outline for extremely useful discussion. Special thanks to James Panton, Tiffany Jenkins and Jacob Reynolds, conversations with whom refined and tested my early thoughts.

The community of scholars of philosophy at Birkbeck College, who have done their best to help me become a better thinker and writer. Fellow members of the Royal Statistical Society, especially the Data Ethics and Governance Section, for helping me clarify what data can and can't, and what it arguably shouldn't, do.

Toby Mundy for seminal early conversations. Ziyad Marar and the organisers of Social Science Foo, and MIT Tech Review, for invitations to their conferences where I met so many interesting people. Radio colleagues, especially Jonathan Brunert on FutureProofing and Divided Nation, Martin Rosenbaum on How To Disagree, and Michael Blastland on The Human Zoo. Brian Tarran, who commissioned pieces for Significance magazine which also fed research for this book. Matt Parker and our team on the show that took me back to Greenham Common.

Holly Marriott at Eighteen O Four, for organising my life so I can do all the other things.

Grahame Willcocks and the other volunteers at Portsmouth Historic Dockyard who were so helpful and hospitable. That I couldn't find room in this book for the full story of my visit to the Block Mills, and the true origins of the assembly line system, is a great sadness to me.

Everyone with whom I've argued history, philosophy, and politics over the years, for forcing me to test my ideas and see other perspectives. I am writing in a language I did not myself make, articulating my thoughts through concepts given to me by others, and I am grateful for that. Those who gave me those ideas, and these habits of thought, may not agree with what I made of them, but that is the nature of human life. I take full responsibility for what I have written.

Index

ONE PLACE. MANY STORIES

Bold, innovative and
empowering publishing.

FOLLOW US ON:

@HQStories

Praise for *Technology Is Not the Problem*

'Great book: a very bold and insightful attempt to grapple with the strangely unexpected social confusion we are experiencing in the 21st century!'
Matt Ridley, author of *How Innovation Works*

'Absolutely captivating . . . Equal parts illuminating and empowering, [this] is essential reading.'
Pete Etchells, author of *Unlocked*

'An urgent, must-read for anyone striving for a nuanced analysis of the complex relationship between technology and society [and] ultimately seeking liberation.'
Tiffany Jenkins, author and broadcaster, presenter of BBC Radio 4's *The History of Secrecy*

'Interesting and informative but also enticing and highly addictive. No, that's not the internet. It's this book. I recommend you read it.'
Stephen Senn, statistician and author of *Dicing with Death*

'A fresh and fascinating perspective. In a world where we are told we are increasingly powerless, Harkness reminds us of our own agency.'
Jenny Kleeman, author of *The Price of Life* and *Sex Robots & Vegan Meat*

'An eloquent and invigorating antidote to amnesia, putting today's technologies in their cultural and historical place – and insisting that the great questions of a digital age remain intransigently human.'
Tom Chatfield, author and tech philosopher

'This is humanism at its best and most uncompromising. Timandra Harkness – principled, courageous, and original – forces us to confront our own ambition and insecurity, arguing that we can only make sense of technology by first knowing ourselves.'
Michael Blastland, broadcaster, originator of BBC Radio 4's *More or Less* and author of *The Tiger That Isn't*

'A riveting analysis of our uneasy relationship with the digital world, and a must-read for anyone who both loves and hates their smartphones.'
Andrew Doyle, author of *The New Puritans*

'If you're tired of pundits blaming all the ills of the world on technology, and crave some fresh thinking, this is the book for you! With her characteristic wit and intelligence, Timandra Harkness provides a balanced and insightful view of how technology is changing our world, packed with unexpected vignettes and sharp observations, that is a pleasure to read.'
Stian Westlake, Economic and Social Research Council Executive Chair